Aquatic Ecosystem of Cold Region

CV CONTENT VIBES
PUBLISHERS • PRINTERS • AGGREGATORS

About the Editors

Dr. Rinkesh Nemichand Wanjari earned his Ph.D. in Fisheries Resource Management from Sher-e-Kashmir University of Agricultural Sciences and Technology (SKUAST), Kashmir, India. He also holds a Master of Fisheries Science (M.F.Sc) in Fisheries Resource Management from the ICAR-Central Institute of Fisheries Education, Mumbai. Currently, he is an Assistant Professor at, the Department of Fisheries Resource Management, School of Fisheries, Centurion University of Technology and Management, Paralakhemundi, Odisha, India. Dr. Wanjari has qualified for the ICAR-JRF (2020), ICAR-NET (2021&2023), and ICAR-SRF (2021). He has published many research papers, abstracts, book chapters, and popular articles. He has specialization in Riverine Fisheries, Fish Stock Assessment, Marine Biology, Trawl Fisheries, Remote Sensing and GIS, Fish and Fisheries, Fisheries Biology, Population Dynamics and Ecological Modelling. He has presented at numerous national and international conferences and participated in various training programs and workshops. His accolades include the Best Poster Presentation at an international conference organized by the University of Mumbai, Best M.F.Sc. Thesis Award, and Best Ph.D. Research Scholar Award.

Mr. Zaheer Abass is pursuing his PhD at Sher-e-Kashmir University of Agricultural Sciences and Technology Kashmir (SKUAST). He also completed his Master of Fisheries Science (M.F.Sc) in Fisheries Resource Management from the ICAR-Central Institute of Fisheries Education, Mumbai. Mr. Zaheer Abass has qualified for the ICAR-JRF (2020), ICAR-NET (2023), and ICAR-SRF (2022). He has been awarded a National Fellowship and Scholarship for Higher Education of ST Students to pursue a Ph.D. degree in 2023. He has his specialization in Fish Taxonomy, Fishing Technology, fish conservation and management, and geospatial tools (QGIS). He has published many research papers, abstracts, book chapters, and popular articles. He has presented at numerous national and international conferences and participated in various training programs and workshops.

Dr. Aadil Hussain Magloo, the editor of this book, holds a Ph.D. in Fisheries Resource Management from the Faculty of Fisheries, SKUAST-K, which is India's only institution dedicated to coldwater fisheries. His expertise lies in the fields of Reproductive biology of fishes, stock assessment, and population dynamics, with a specific focus on coldwater species found in the Kashmir Valley. His contributions to the field are backed by a solid research portfolio, with multiple publications addressing key issues in fish biology, stock assessment, and the sustainable management of coldwater fisheries. Dr. Magloo is also the holder of a patent for a smart fishing rod design, showcasing his innovative approach to practical challenges in fisheries. Dr. Magloo has actively participated in various national and international conferences, and his academic excellence has been recognized with several accolades, including the prestigious Best Thesis Award. His practical experience extends to working as an assistant manager at K2 Aquaculture in Kangan, Ganderbal, Kashmir, which is a state-of-the-art recirculatory aquaculture project with an impressive production capacity of 1,800 tons of trout annually. His hands-on expertise covers fish handling, raising, and growing from the egg stage to market size, further showcasing his comprehensive knowledge of aquaculture.

Professor (Dr.) Tasaduq H. Shah is currently working as Professor cum Chief Scientist of Fishery Biology in the Division of Fisheries Resource Management, Faculty of Fisheries, SKUAST-Kashmir, Rangil,Ganderbal (J&K). He has earlier served as Head, Division of Fisheries Resource Management, Faculty of Fisheries, SKUAST-Kashmir and has M.F.Sc. and Ph.D. degree from Central Institute of Fisheries Education (ICAR-CIFE), Mumbai. He specialises in fish biology, biodiversity, fish stock assessment, conservation and management. He is the recipient of Hiralal Chaudhuri Gold Medal for merit during his master's programme and has twenty-seven years of experience in fisheries science. He has guided dozens of masters and doctoral students and published 95 research publications in leading national and international scientific journals.

Professor (Dr.) Farooz Ahmad Bhat Dean, Head, and Professor-cum-Senior Scientist at Sher-e-Kashmir University of Agricultural Sciences and Technology (SKUAST), Kashmir, specializing in Fish Endocrinology and Reproductive Physiology at the Division of Fisheries Resource Management (FRM). Holds an M.Phil. and Ph.D. in the field. The recipient of numerous awards and honors, including recognition as a Functional Area Expert (FAE) in Ecology and Biodiversity (NABET Accredited) and Fellowships from the Coldwater Fisheries Society of India (2019) and The Society of Life Sciences (2019). Awarded the Best Teacher Award at the university level for the academic year 2017-18. Possesses over 21 years of teaching experience, instructing B.F.Sc., M.F.Sc., and Ph.D. students since 2007, and over 19 years of research experience in Fisheries science.

Aquatic Ecosystem of Cold Region

Rinkesh Nemichand Wanjari
Division of Fisheries Resource Management, Faculty of Fisheries
Sher-e-Kashmir University of Agricultural Science and Technology
Kashmir, Ganderbal, India-190025

Zaheer Abass
Ph.D. Research Scholar
Division of Fisheries Resource Management, Faculty of Fisheries
Sher-e Kashmir University of Agricultural Science and Technology
Kashmir, Ganderbal, India-190025

Aadil Hussain Magloo
Ph.D. Research Scholar
Division of Fisheries Resource Management, Faculty of Fisheries
Sher-e-Kashmir University of Agricultural Science and Technology
Kashmir, Ganderbal, India-190025

Tasaduq H. Shah
Ph.D. Fisheries Resource Management
Professor cum Chief Scientist at
Sher-e-Kashmir University of Agricultural Sciences and Technology
Kashmir, Ganderbal, India-190025

Farooz Ahmad Bhat
Professor cum Chief Scientist, Dean Faculty of Fisheries
Sher-e-Kashmir University of Agricultural Science and Technology
Kashmir, Ganderbal, India-190025

CV CONTENT VIBES
PUBLISHERS • PRINTERS • AGGREGATORS

PUBLISHERS • PRINTERS • AGGREGATORS

1168, Sector 13, Urban Estate
Karnal-132 001, Haryana
Tel: 91-84470 75807, 18440 41168
Email: contentvibesppa@gmail.com
www.contentvibes.in

Print ISBN: 978-93-6755-647-4

eISBN: 978-93-6755-419-7

भा.कृ.अनु.प.- केन्द्रीय मात्स्यिकी शिक्षा संस्थान
ICAR- CENTRAL INSTITUTE OF FISHRIES EDUCATION
(A University Established Under Sec. 3 of UGC Act 1956)
Ministry of Agriculture & Farmers Welfare, Govt. of India

Dr. Karankumar K. Ramteke
Scientist
(Fisheries Resource Management)

Foreword

The Kashmir Valley is a true tapestry of natural beauty, renowned for its breathtaking landscapes, serene waters, and rich biodiversity. Among its many treasures, the aquatic resources and fisheries of the region stand out as a vital component of both the environment and the livelihood of its people. *Aquatic Ecosystem of Cold Region* is a pioneering work that brilliantly captures the complexity and richness of this aquatic ecosystem.

This book takes readers on a journey through the diverse fish species of the valley, exploring not only their biology but also their importance to local communities and the wider economy. The authors present a detailed and insightful examination of the intricate relationships between the region's fisheries and the challenges posed by environmental changes, particularly climate change. What makes this work truly unique is its forward-thinking approach, offering practical strategies for sustainable fisheries management and conservation.

In a time of growing ecological pressures, this book stands as an essential resource for researchers, policymakers, and environmentalists alike. Its comprehensive analysis, combined with a hopeful outlook on the future of Kashmir's fisheries, ensures that this work will be a cornerstone for future efforts to protect and promote the region's natural heritage.

As you delve into *Aquatic Ecosystem of Cold Region*, I am confident that you will gain a deeper appreciation for the invaluable role that fisheries play in this remarkable region and the urgency with which we must act to preserve them for generations to come.

(Karankumar K. Ramteke)

पंच मार्ग, ऑफ यारी रोड, वरसोवा, अंधेरी (प), मुंबई-400 061. (भारत)
Panch Marg. Off Yari Road, Versova, Andheri (W), Mumbai-400 061. (India)

कार्यालय/Office : 022-26361446/7/8
Fax: 022-26361573
Website: www.cife.edu.in

Perface

"Aquatic Ecosystem of Cold Region" is a comprehensive exploration of the diverse and dynamic aquatic resources that define the unique landscape of the Kashmir region. This book delves into the rich tapestry of fish species, their habitats, and the intricate relationship between the aquatic ecosystems and the local communities that have relied on them for generations. Through sixteen captivating chapters, the reader is taken on a journey that unravels the complex web of factors shaping the fisheries sector in Kashmir. From the biology and ecology of iconic species like the Schizothoracinae and Salmonid fishes to the impact of climate change and the introduction of exotic species, this book offers a well-rounded understanding of the challenges and adaptations necessary for the sustainable management of these precious resources. The book also sheds light on the traditional fishing practices, gear, and crafts that have been passed down through the ages, providing a glimpse into the rich cultural heritage that is inextricably linked to the aquatic life of the Kashmir Valley. Furthermore, the exploration of sports fisheries development and the associated socio-economic benefits underscores the multifaceted role of fisheries in the region. Ultimately, "Aquatic Ecosystem of Cold Region" is a testament to the remarkable diversity and resilience of the aquatic ecosystems in this trans-Himalayan region. It serves as a valuable resource for researchers, policymakers, and all those who are invested in preserving the delicate balance between human activities and the natural world. Through its forward-looking perspective on sustainable management and conservation strategies, this book paves the way for a future where the fisheries of Kashmir can thrive in harmony with the needs of the local communities and the ever-changing environmental landscape.

Acknowledgements

The editors would like to express their heartfelt gratitude to all the contributing authors for their invaluable chapters in this comprehensive book on the fisheries of the Kashmir Valley. The collective expertise and research shared by these authors have been instrumental in creating a thorough and insightful exploration of the aquatic resources, fish diversity, and the unique challenges and opportunities facing the fisheries sector in this trans-Himalayan region. This book would not have been possible without their tireless efforts and commitment to advancing the understanding of this important field. The editors also would like to express their appreciation to the local communities, fishermen, and stakeholders in the Kashmir Valley, whose invaluable knowledge, experiences, and perspectives have greatly enriched the content of this book. Their insights and contributions have been instrumental in capturing the unique and nuanced aspects of the fisheries in this remarkable region.

Contents

1

Introduction to Fish and Fisheries of Kashmir

Tabish Farooq*, Farooz A. Bhat, Durdana Qazi, Inab Majeed Bala, Saima Jan and Zaheer Abass

Division of Fisheries Resource Management, Faculty of Fisheries SKUAST K Rangil, Ganderbal, Jammu and Kashmir

Abstract

This chapter "Introduction to fish and fisheries of Kashmir" provides a comprehensive overview of the region's rich aquatic biodiversity and its significance to local communities. It explores the diverse species of fish native to Kashmir, emphasizing their ecological, economic, and cultural importance. The chapter delves into the history and development of fisheries in the region, highlighting traditional fishing practices and the evolution of modern aquaculture techniques. Key challenges facing the fisheries sector, such as environmental threats, overfishing, and climate change, are examined alongside current conservation and management strategies. This introduction sets the stage for a deeper understanding of the intricate relationship between Kashmir's saquatic ecosystems and its people, offering valuable insights for sustainable fisheries development.

Keywords: *Aquatic biodiversity, Climate change, Traditional fishing, Sustainable fisheries, Kashmir*

Introduction

Fish and fisheries play a significant role in Kashmir's economy and people's livelihoods. Kashmir abounds with water bodies the lakes, rivers, and streams run the course to produce a favourable environment for numerous species of fish. Apart from the natural water bodies, fish farming has also emerged significantly in Kashmir, where several small-scale fish farmers are engaged in the fish rearing and fisheries activity in ponds and race ways, both for commercial and subsistence purposes (Islam, 2020). The fish farmers do more than just contribute to local food supplies; they also provide money and work for people locally. However, the fish and fisheries sector of Kashmir is faced with various challenges. The challenges range from overfishing, environmental degradation, and pollution to the lack of infrastructure for fish handling and processing and climatic change. Despite all odds against this sector, the government and other agencies/organizations are trying their best to achieve sustainable management and development of fisheries in Kashmir. Moreover, full- and part-time work by fish farmers and fish workers, seasonal or occasional, supplement thousands of livelihoods with additional income in the sector's activities (Qayoom and Bhat, 2015). Most of those engaged in small-scale fisheries and aquaculture in Kashmir are self-employed, hence directly contributing towards food security within their households and communities. Due to the ever-increasing demand for protein-rich food, the production of fish has elevated around the world. According to the Food and Agriculture Organization, fish production attained 179.2 million metric tons in 2020, which is valued at USD 345 billion (FAO, 2020). This shows a large increase from the years of the 1990s when world fish production was 100 million metric tons.

Fish Production of India

Trends in fish production of India indicate a continuous rise of an annual average growth rate of 8.61% over the period from 2014-15 to 2021-22 (dof.gov.in). It is due to inland fisheries, which contributes around 75 percent of the total fish production in the year 2021-22, while the remaining 25 percent is from marine fisheries (dof.gov.in). Value-wise, the country's fish production has also had a massive outflow in exports, touching US$ 8.09 billion in FY 2022-23 (Handbook of fisheries statistics, 2023).

Key Drivers of Growth

Several factors are behind the rapid growth of fish production in India. First, the growing demand for fish emanates from the growing population coupled with the aspiration of the people toward protein-rich sources of food. Second, India has diverse water resources rich in biodiversity, which imparts great potential for fish production. Government initiatives like the Pradhan Mantri Matsya Sampada Yojana and Pradhan Mantri Matsya Kisan Samridhi Sah-Yojana are centered on hiking the earnings and incomes of fishermen, fish vendors, and the micro & small enterprises engaged in this sector.

Fish production of Kashmir

Kashmir, a vale of beauty, is situated in the northern part of the Indian sub-continent. This magnificent land of beauty is well-known for its tremendous scenery, snow-capped peaks, and crystal-clear lakes. This State is gifted with natural water resources that cover about 0.40 lacs ha in total water spread area. These exist in the shape of torrential coldwater streams, lakes, rivers, springs, and high-altitude lakes. Besides the incredible landscapes and serene lakes, the Kashmir Valley, embedded within the grand Himalayan ranges, boasts great biodiversity. One prominent part of this biodiversity is the wide variety of fishes residing in the lakes, rivers, and streams of the Kashmir Valley. These fishes play a crucial role in the ecosystem here and form a part of cultural and economic heritage for centuries. Various species of fishes inhabited Kashmir Valley, all having their characteristics and adaptations. Among them, the family Cyprinidae recorded the highest species richness. The fish diversity of the Kashmir Valley plays a pivotal role not only in ecological but also in economic aspects. The indigenous fishes of the Kashmir Valley form an important source of livelihood among many local communities through various activities in fishing and fish farming.

The fish production in Kashmir has shown a mixed trend over the years. While the overall fish production in Jammu and Kashmir State has increased significantly from 95,418 quintals in 1980-81 to 200,800 quintals in 2015-16, the production in Kashmir province has exhibited an undesirable decline in recent decades. The production of *Schizothorax richardsonii*, an indigenous fish species to Kashmir, went down from 2.5 thousand quintals in 1980-81 to just 0.6 thousand quintals in 2012-13. Their decline was more pronounced in Dal Lake compared to the freshwater Wular Lake, and was attributed to stocking with exotic species, eutrophication of their habitats, and widespread encroachment.

The fish production in Kashmir province was steadily increasing until 2010-11, but has remained almost stagnant between 1998-99 to 2010-11. Surprisingly, there has been a decline in fish production in Kashmir since then, which may be due to the degradation of fish habitats.The growth rate estimates revealed that while the overall fish production in Jammu and Kashmir state had a significant growth from 1980 to 2015, the growth was more pronounced in Jammu province compared to Kashmir. The trout exhibited a significant desirable trend towards growth, but the growth in mirror carp and country fish was lower (Baba *et al.*, 2019).

The overall fish production in Jammu And Kashmir UT has increased from 95,418 quintals in 1980-81 to 200,800 quintals in 2015-16 (JK fisheries department, 2016). This represents a substantial growth in fish production over the past few decades. The growth rate analysis reveals that the compound annual growth rate (CGR) of total fish production in the UT was 2.31% from 1980-81 to 2015-16, which is statistically significant (Baba *et al.*, 2015). This indicates that the UT has made considerable progress in expanding its fish production capabilities.

Share of Jammu and Kashmir in Indian Fish Production

Fish production in Kashmir Province, J&K depicted in (Fig. 1). However, as shown in Figure 2, the share of Jammu and Kashmir in the total fish production of India has declined from about 0.39% during 1980-81 to 0.21% in 2015-16 (Baba *et al.*, 2019). This suggests that the growth in fish production in Jammu and Kashmir has been slower compared to the overall growth in fish production at the national level. Fish caught in Jammu & Kashmir and trout production in UT of Jammu and Kashmir are presented in (Table 1 and 2). Figure 3 showing increasing trend in trout production and revenue generation in Kashmir.

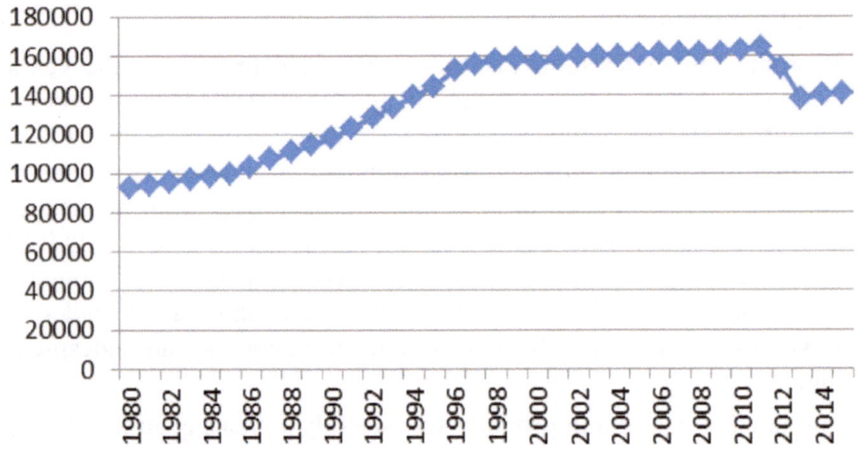

Fig. 1: Fish production in Kashmir Province, J&K (Baba *et al.*, 2019)

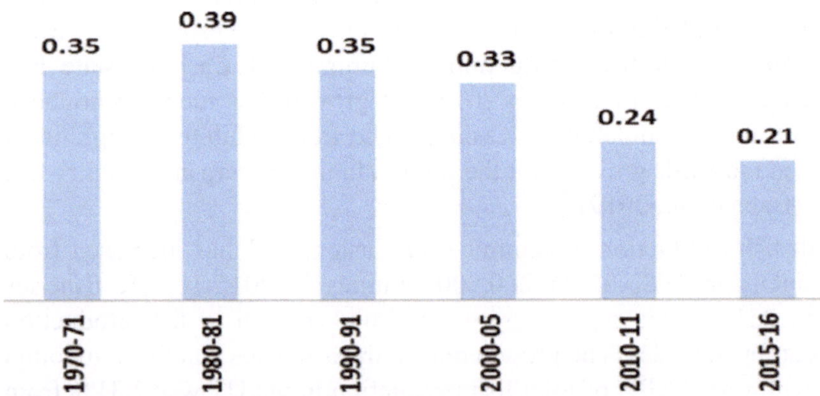

Fig. 2: Share of J&K in Indian fish production (%)

Table 1: Fish caught in Jammu & Kashmir (quintals) (Baba *et al.*, 2019)

Year	Kashmir				Jammu	State
	Trout	Mirror carp	Country fish	Total		
1980-81	79	62712	29914	92705	2713	95418
1990-91	62	82558	35391	118011	17000	135011
2000-01	792	114323	41295	156410	28257	184667
2010-11	1950	117413	43395	162758	34242	197000
2015-16	2980	101104	36670	137774	63026	200800
CGR (%)	12.11*	1.79*	0.93*	1.59*	7.55*	2.31*
	(0.67)	(0.20)	(0.13)	(0.17)	(0.54)	(0.15)

Table 2: Trout production in UT of Jammu and Kashmir (JK fisheries department, 2023)

Year	Production/biomass (Tonnes)	Revenue (Lacs)	Private units	Hatcheries
2019-20	650	360.16	534	0
2020-21	650	403.13	718	0
2021-22	1663	422.79	936	1
2022-23	1990	427.37	1144	3

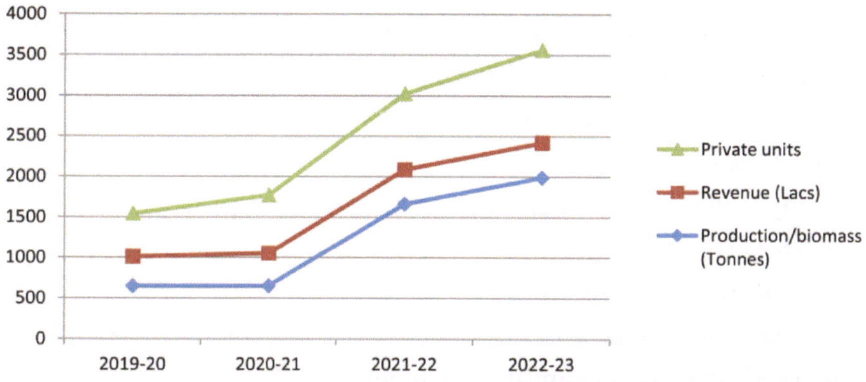

Fig. 3: Graph showing increasing trend in trout production and revenue generation in Kashmir.

Icthyofaunal Diversity of Kashmir

Kashmir's fish species include Cyprinidae, Cobitidae, Siluridae, Salmonidae, and Poecilidae, with the former dominating the valley's water systems. Cyprinidae is dominated by schizothoracid fishes. Snow trouts, an indigenous fish species, are highly sought after on the market.

List of fish species reported from the Kashmir Valley (Bhat *et al.*, 2020).

Order Cypriniformes (Family Cyprinidae)

1. *Schizothorax niger* Heckel
2. *S. esocinus* Heckel
3. *S. labiatus* McClelland
4. *S. curvifrons* Heckel
5. *S. plagiostomus* Heckel
6. *Cyprinus carpio* var. *communis* Linnaeus and *C. c.* var. *specularis* Linnaeus
7. *Carassius carassius* Linnaeus
8. *Puntius conchonius* Hamilton
9. *Bangana diplostoma* Heckel
10. *Crossocheilus diplochilus* Heckel
11. *Diptychus maculatus* Steindachner
12. *Tor putitora* Hamilton
13. *Hypophthalmichthys molitrix* Valenciennesa
14. *Ctenopharyngodon idella* Valenciennesa

    Family Cobitidae (Subfamily Botinae)

15. *Botia birdi* Chaudhari

    Family Balitoridae (Subfamily Nemacheilinae)

16. *Schistura punjabensis* Hora
17. *Triplophysa marmorata* Heckel
18. *T. kashmirensis* Hora

    Order Siluriformes (Family Sisoridae)

19. *Glyptosternon reticulatum* McClelland
20. *Glyptothorax kashmirensis* Hora

    Order Salmoniformes (Family Salmonidae)

21. *Salmo trutta fario* Linnaeus
22. *Oncorhynchus mykiss* Walbauma

    Order Cyprinodontiformes (Family Poeciliidae)

23. *Gambusia holbrooki* Girard

Heckel's (1838) account is the oldest on fishes in Kashmir; he listed sixteen (16) species of fish from the valley, thirteen (13) of which were members of the Cyprinidae family. Several reports on the Ichthyofauna of the Valley have been made since Heckel 1838 (Day 1876; Silas 1960; Das and Subla 1964; Das 1965; Yousuf 1996; Kullander *et al.*, 1999; Enderlin and Yousuf 1999; Balkhi 2007). The number of fishes reported by these researchers varies from 18 to 42. This discrepancy in the number of fish species appears to have resulted from identification uncertainty or synonymy. Kullander *et al.* (1999) synonymized numerous species, resulting in a lower number of fish species reported in previous research. Balkhi (2007) reported forty four (44) species of fish from Kashmir, however Yousuf (1996) recorded forty two (42) species. But according to Kullander *et al.* (1999) and Bhat *et al.*, (2010), there are now just twenty-two (22) species of fish in the Kashmir region. Out of 23 fin fish species reported from Kashmir 19 are cold water species and 4 are warm water species.

History

The history of fishing and fisheries in Kashmir is deeply related to the cultural heritage and natural resources of the state. Kashmir's angling history is as old as 1900 when both brown rainbow trout were introduced by the British colonialists to its rivers. The case of trout farming in Kashmir is particularly interesting. In 1899, the first lot of 10,000 trout eggs arrived from the United Kingdom. These were sent as a gift by the Duke of Bedford to the Maharaja of Jammu and Kashmir. This lot naturally perished during transport due to lack of transport facilities. It was the second successful consignment of trout ova from Scotland in the year 1900 that opened doors to trout farming in Kashmir. Part of the fry was then shifted to the locations in and around Srinagar and a part of them was reared at the premises of a private carpet factory owner in the centre of the city. From then on, the growth of trout farming began slowly but steadily and became prominent and adaptive to Kashmir (JKfisheries.in). These varieties of trout had their habitat to great advantage in the cool, highly oxygenated waters that emanated from the Himalayas and turn Kashmir into an angler's Eldorado, drawing enthusiasts from each corner of the globe. The streams and lakes of Jammu and Kashmir-especially the Sind and Lidder Rivers, the Gangbal Lake, and other tributaries of the River Jhelum-are believed to be at their best with brown and rainbow trout, offering an angler paradise for various fishing experiences. The geographical diversity of the region with pure mountain air, serene landscapes, and clear lakes provides an ideal setting for trout fishing and creates an once-in-a-lifetime experience for fishing enthusiasts. Besides trout, there are some indigenous fish species in the rivers of Kashmir which can add to the angling experience of catching and witnessing the fighting spirit among such fishes.

Fisheries have also held an important place in the region historically. So far as the contribution to trout production is concerned, it has been immense in the Indian subcontinent, particularly in the Kashmir valley. The mode of fishing reflects a blend with tradition and modernity whereby locals who flyfish appear to use gear and techniques dating 75 years back to British fishing tackle. Rivers in Kashmir are further divided into beats managed by a beat guard and a ghillie, maintaining a system akin to mid-20th century England, whereby anglers book and pay for fishing specific sections of the river (JKfisheries.in). Though occupation-wise, fishing is very significant, carrying rich historical-cultural values in Kashmir; still, it has received some attention in the recent past with regard to the livelihood of fishermen. These programs aim at empowering local fishing communities, drumming home the message of sustainable fisheries management, and act as a harbinger of conserving aquatic resources of Kashmir for posterity. Fisheries have a long and fascinating history in the Kashmir region, playing a significant role in the local culture and economy.

Traditional Fishing Practices

Kashmir has adequate natural water resources in the form of rivers like Jhelum and its tributaries, and famous lakes like Dal, Wular, Manasbal, and Anchar. These water bodies supported the habitat for a variety of fish species, both native ones and exotic ones like trout. The traditional fishing practices involves the use of traditional crafts and gears. The craft usually employed for fishing is made up of deodar wood that is not affected by water. The gears include cast net, locally called as zaal, gill net, locally called shaitan zaal which is now banned in Kashmir waters, long line, locally called as wael raaz, panzer and narchoo that are spear headed wooden gears that are also now banned but are still used by artisanal fisherman and pole and line commonly called as bislai and usually used by anglers to catch game fishes (Syed *et al.*, 2020).

Fisheries Management and Development

From sport fisheries in Kashmir, fisheries management has come a long way to become a holistic package of conservation, infrastructure development, and introduction of modern fish farming technologies. Trout farming has increased considerably in Kashmir over the years. The Brown Trout, which grows in the waters of Kashmir, caters to a sizeable number of anglers and is an important ingredient for the tourism industry. On the other hand, Rainbow Trout is grown for commercial purposes.

Historical Evolution: The Department of Fisheries in Kashmir came into existence in 1903 as the Department of "Game Preservation" following the successful establishment of trout in this valley. Therefore, up to 1978, the department remained confined mostly to sport fisheries and did practically nothing in the sphere of conservation phase viewing the ramification at the infrastructure, trained manpower, and modern fish farming technology. However, in 1978, the department was reorganized into establishing district-level offices, and detailed surveys were conducted to identify potential areas under fisheries development (JKfisheries.in).

Infrastructural Development: The Department of Fisheries in Kashmir, subsequent to its reorganization, developed a good number of carp and trout farms, hatcheries, rearing units, sale centers, laboratories, angler lodges, and aquarium-cum-awareness centers in every one of the three areas of the state. This was supposed to be for increasing fish production and conserving fish resources of the state as well as promoting sustainable fisheries practices.

Involvement of Private Sector: The department introduced fish culture in the private sector under various schemes sponsored by the state and central governments to mitigate unemployment among educated youth. It was aimed at creating an avenue for employment generation in the fisheries sector and to promote self-enterprising skills through this scheme.

Technological Advances: In recent past years, there has been a shifting process towards adopting modern fish farming technologies and practices to increase production efficiency and meet the increasing regional fish demand. The development of trout rearing units, RAS trout units, hatcheries, and feed mills equipped with advanced technology played a vital role in enhancing fish production and quality in Kashmir. In nutshell, fisheries management in Kashmir has traversed a long distance from mere sport fisheries to a much more holistic approach of conservation-based infrastructure development, Holts, private sector participation, and adoption of modern fish farming technologies that not only increased the fish production but also contributed much towards economic development and sustainability of fisheries in the UT. Despite these efforts, the fisheries sector in Jammu and Kashmir has contributed less than 2% to the agricultural economy, and its contribution to the state's net domestic product has been meager (Qayoom and Bhat, 2015). To enhance the growth and contribution of this sector, there is a need to explore the reasons responsible for the negative growth in mirror carp and country fishes, maintain a balanced stock of local and exotic species, adopt scientific recommendations in fish culture, and focus on value addition. Preservation of water bodies, removal of encroachment, and regulation of fish marketing in the state are also crucial for the sector's development.

Economic and cultural significance

Not only has fish farming changed the lives of individual farmers, but it has also come up with a solution regarding unemployment in this region. Economic benefits associated with the growth of a sustainable and widespread trout fish industry in Kashmir are quite large, offering means of livelihood as well as food security to people living in the hills (Sheikh and Khan, 2019). Fisheries have also been very much associated with the cultural heritage of Kashmir. The serene valleys and pristine water bodies of Kashmir provide an ideal environment for trout fish farming, and it is pride and identity accumulating among erstwhile lots of people living here. From the traditional methods of fishing to the introduction and development of trout farming, the history of fisheries stretches a pretty long way in Kashmir. The industry evolved over time into a mainstay of the economy and culture of Kashmir. The fisheries were part of the mainstay of livelihoods and hence food security in Kashmir.

Scope of Fish and Fisheries in Kashmir Valley

The Kashmir Valley, nestled in the northern part of India, is renowned for its breathtaking landscapes, pristine water bodies, and unique biodiversity. Among its various natural resources, the fish and fisheries sector holds significant potential. This essay explores the scope of fish and fisheries in the Kashmir Valley, highlighting the region's unique conditions, current practices, potential for growth, and the challenges it faces.

Unique Conditions Favoring Fisheries in Kashmir Valley

Kashmir Valley abounds in freshwaters, having rivers, lakes, streams, and springs. These impoundments offer a congenial environment for natural fish populations and aquaculture. Some of the main features that make this region very ideal for fisheries include:

1. **Cold Water Temperature**: The valley has cold waters due to its cold climate; these are fast-flowing and of high clarity, creating optimal conditions for keeping cold-water fish species, specifically rainbow and brown trout, which are the main fish species farmed in the valley.
2. **Rich Biodiversity:** The habitat is the abode of various species of indigenous fish including Schizothorax (snow trout) well-adjusted to local environmental conditions. The biodiversity of this habitat with regard to aquatic fauna translates into a potential for a diversified and maintainable fisheries sector in Kashmir (Pandit and Qadri, 2014).
3. **Natural Springs and Streams: Several** natural springs and streams provide high-quality water in optimal conditions for fish farming, hence healthy and productive aquaculture.

Current Practices in Kashmir's Fisheries Sector

Kashmir fisheries can basically be divided between capture fisheries and aquaculture. Both have their own characteristics and practices:

1. **Capture Fisheries:** In Kashmir, the capture fisheries are connected with the natural harvesting of fish from rivers, lakes, and several other water bodies. These mainly include fisheries from the Dal Lake and Wular Lake, both having varieties of native fish species.
2. **Aquaculture:** In Kashmir, the aquaculture industry is dominated by culture for rainbow trout and carps. The initial introductions related to the farming of trout date back to the beginning of the 20th century, with more extensive developments during the last few decades. Hatcheries and farms are functioning throughout the Valley to assist in the breeding and production of trout.
3. **Governmental efforts:** These involve a set of initiatives by the Government of Jammu and Kashmir with the view to promote fisheries and aquaculture. The major initiatives are establishing trout hatcheries, giving subsidies, providing financial assistance to fish farmers, and training programs to enhance skills and upgrade knowledge.

Potential for Growth in Kashmir's Fisheries Sector

Kashmir does have a colossal potential for growth in the fisheries sector, and it could overflow from many factors.

1. **High-Value Species:** Trout is a high-value species that assures very good economic returns. Thus, increasing trout farming with enhancement of its production efficiency would increase the income of local farmers and benefit the economy of the region as a whole.
2. **Ecotourism:** Kashmir has beautiful landscapes and serene water bodies that attract tourists from around the world. A mix of fisheries into ecotourism can provide unique opportunities in the form of angling, fish farm tours, etc., enhancing both the sectors.
3. **Diversification:** While core production will be in trout farming, there is a scope for diversification into other cold-water species and even introducing warm-water species in controlled environments. This will decrease risk while increasing the overall production.
4. **Export Market:** With proper infrastructure and quality control, the products of Kashmir fisheries, in particular trout, have the potential to reach international markets. This will require efforts to meet global standards with robust, associated supply chains (Ahmad and Reshi, 2017).

Challenges Faced by the Fisheries Sector

Although the fisheries sector in Kashmir holds immense potential, several bottlenecks need to be overcome:

1. **Political Instability:** The entrenching political instability in the region has had its toll on Infrastructure development and access to markets. Ensuring a stable environment is sine qua non to sustaining growth.
2. **Environmental degradation and loss of breeding grounds:** Huge threats to aquatic ecosystems exist due to pollution, habitat destruction, and climate change. Such conservation and restoration will be important for the long-term sustainability of the fisheries sector.
3. **The insufficiency of infrastructure facilities**, such as cold storage, transport network, and a fleet of modern farm equipment, acts as a constraint on the progress of the sector. Necessary investment in the infrastructure shall help raise production and marketability. Unorganized markets and unregulated marketing practices also add to this scenario.
4. **Training and Education:** The majority of local fish farmers are not given adequate exposure to modern techniques and knowledge. Extension of training and education activities will improve farming practices and enhance the general standards.
5. **Introduction of exotic species and unbalanced stocking:** The introductions of exotic fish species and unbalanced stocking practices have had a significant negative impact on the indigenous fish species in Kashmir's water bodies. The introduction of fast-growing exotic fish species, such as carp, has led to the encroachment of natural water bodies and the decline of native Schizothoracid (snow trout) species in Kashmir.
6. **Strong social taboos:** the fishing community in Kashmir lives in abject poverty, with low education levels and lack of access to modern fishing technologies and government support programs. Many fishermen have had to abandon fishing as their primary livelihood due to the decline in fish stocks, and have taken up other jobs, further threatening the traditional fishing way of life.

References

Ahmed, I., & Reshi, Q. (2017). Trout Farming in Kashmir: An Overview. Journal of Fisheries and Aquaculture Research, 12(3), 45-56.

Baba SH, Wani MH, Wani S.A. (2015). Pesticide application on apple: Issues of marketing and ecological environment. Policy paper, Kashmir and Ladakh Agri-watch Centre. Directorate of Research, Policy Research Series DR/SKUAST-K/2015/10, SKUAST-K. www.skuas tkashmir.ac.in.

Balkhi MH (2007) Fish diversity of Jammu & Kashmir and its conservation. In: Patloo (ed) Kashmir speaks. G. M. Publishers, Kashmir, pp 104–118.

Bhat F. A., Yousuf A. R. and. Balkhi M. H (2020). Diversity of Fishes in Jammu and Kashmir State. In: Ghulam Hassan Dar and Anzar A. Khuroo (Ed.) Biodiversity of the Himalaya: Jammu and Kashmir State.

Bhat FA, Yousuf AR, Balkhi MH, Mahdi MD, Shah FA (2010) Length-weight relationship and morphometric characteristics of Schizothorax spp. in the River Lidder of Kashmir. Indian J Fish 57:73–76

Das SM (1965) A revision of the fish species inhabiting Kashmir province. Kashmir Sci 2(1–2):13–19

Das SM, Subla BA (1964). The Ichthyofauna of Kashmir, Part II. The speciation of Kashmir fishes.Ichthyologica 3:57–62

Day F (1876). On the fishes of Yarkand. Proc Zoology Soc London: 781–807.

Department of Fisheries, Jammu and Kashmir Government. (2021). Annual Report on Fisheries.

Department of Fisheries, Jammu and Kashmir Government. (2023). Annual Report on Fisheries.

Enderlin O, Yousuf AR (1999). The environmental impacts f the Uri hydropower project on the fish community in the river Jhelum. In: Nyman L (ed) River Jhelum, Kashmir valley: impacts on the aquatic environment. SWEDMAR, Göteborg, pp 169–186

Environment and Energy. Kashmir University, Srinagar, pp 75–120. Environment SWEDMAR, Göteborg.

Environment SWEDMAR, Göteborg.

Handbook on Fisheries Statistics, (2023). Department of Fisheries Ministry of Fisheries, Animal Husbandry and Dairying Government of India, New Delhi.

Heckel JJ (1838) Fische aus Caschmir. Karl Freiherrn von Hugel. Wien. Gedruckt Bei Den P. P.

Kullander, S.O., Fang. F., Delling, B. and Ahlander, E. (1999). The fishes of the Kashmir Valley.

Pandit, A. K., & Qadri, M. Y. (2014). Aquatic Biodiversity and Fishery Potential of Kashmir Himalayas. International Journal of Environmental Sciences, 9(2), 101-115.

pp. 99-162. In: Lennart Nyman (Ed.), River Jhelum, Kashmir Valley, impact on the aquatic

Qayoom, I., & Bhat, B. A. (2015). Change in Fishing Patterns in Jammu and Kashmir Provinces of J and K-A Comparative Study. Current World Environment, 10(1), 179.

Rather Tajamul Islam (2020). Fisheries Sector In Jammu and Kashmir—Growth and Perspectives. International Journal of All Research Writings, 1(7), 1-7.

Sheikh, B. A., & Khan, M. A. (2019). The Role of Fisheries in Livelihood Security in Kashmir Valley. Journal of Rural Development, 38(1), 75-90.

Syed, Nimat & Mohite, Ashish & Sadawarte, Rahul & Desai, Ajay & Shah, Tasaduq (2020). Design aspects of fishing crafts and gears of Wular Lake of Kashmir, India.

Yousuf AR (1996) Fishery resource of Kashmir. In: Khan AH, Pandit AK (eds) Ecology, Environment and Energy. Kashmir University, Srinagar, pp 75–120.

2

Aquatic Resources of the Trans-Himalayan Region

Dechen Chuskit, Farooz Ahmad Bhat, Rinkesh Nemichand Wanjari, Hudisa Banoo* and Zaheer Abass

Division of Fisheries Resource Management, SKUAST-Kashmir, Jammu and Kashmir

Abstract

The Trans-Himalayan region, often known as the Tibetan Plateau, is a large, high-altitude region in Central Asia that includes sections of Tibet, Ladakh, and the Tibetan Autonomous Region. The vast ice fields and glaciers of this region, also referred to as the "Third Pole," are vital to the world's freshwater resources because they feed significant rivers like the Indus, Brahmaputra, and Yangtze. The region's distinctive geography which is made up of high peaks, wide plateaus, and deep valleys as well as its severe climate conditions all contribute to its rich biodiversity and intricate weather patterns. Specialised plants and animals, such as the snow leopard and Tibetan antelope, can be found in the region's ecosystems because they have evolved to harsh environments. Because they depend on freshwater environments for both agriculture and cultural activities, indigenous tribes have established traditional livelihood patterns that are strongly linked to the environment. Important aquatic environments sustain biodiversity and control temperature, including high-altitude lakes like Pangong Tso and Tso Moriri. The Trans-Himalayan aquatic resources are critically threatened by pollution, habitat fragmentation, overexploitation, and climate change, notwithstanding their biological significance. Aquatic ecosystems and human settlements are being impacted by changes in precipitation patterns brought on by global warming and glacial retreat, which are reducing water supply and quality. There are now conservation initiatives in place with an emphasis on community involvement, sustainable management, and habitat restoration. In order to preserve these essential ecosystems for future generations, prospects for sustainable management in the future include strengthening climate resilience, combining traditional knowledge with scientific methods, and promoting international collaboration.

Keywords: *Trans-Himalayan region, Tibetan Plateau, High-altitude, Central Asia biodiversity, Climate change, Pollution*

Introduction

The Trans-Himalayan region, often known as the Tibetan Plateau, is a vast region of Central Asia that is covered with high altitude topography. This enormous and untamed terrain, with its lofty peaks, wide plateaus, and profound valleys, is fundamental in determining the biodiversity and climatic patterns of the surrounding area. Geographically speaking, it includes portions of the Tibetan Autonomous Region (TAR), Ladakh (India), and Tibet (China), among other places, making it a varied and culturally rich region. The Trans-Himalayan region, with an average height of 4,500 metres (14,800 feet), is sometimes referred to as the "Third Pole" because of its large ice fields and glaciers, which hold enormous amounts of freshwater that are essential for areas downstream. Major rivers like the Indus, Brahmaputra, and Yangtze, which supply water to millions of people in South and East Asia, are fed by these glaciers (Xu *et al.*, 2009). The harsh climatic conditions in the area present special possibilities as well as problems for the local population and fauna. Extreme weather, such as low oxygen levels and frigid temperatures, have prompted the emergence of specialised plants and animals that are suited to these high-altitude environments. For example, the Tibetan Plateau is home to famous animals that flourish in its harsh environment, including the Tibetan antelope (*Pantholops hodgsonii*) and the snow leopard (*Panthera uncia*) (Zhao and Guo, 2020). Furthermore, indigenous people have established traditional livelihood practices tailored to the severe circumstances of the Trans-Himalayan area, which carries considerable cultural and spiritual significance for them. These communities depend on the freshwater ecosystems and other natural resources of the area for agriculture, subsistence farming, and cultural activities (Immerzeel and Droogers, 2018).

Geography and Climate of the Trans-Himalayas

The Trans-Himalayan region is a large high-altitude territory situated north of the Greater Himalayas. It is also known as the Tibetan Plateau or the Roof of the World. At an average elevation of 4,500 metres above sea level, this plateau, which spans more than 2.5 million square kilometres, is among the tallest and biggest in the globe (Shi *et al.*, 2019). This region is distinguished by its vast plains, mountain ranges like as the Kunlun, Karakoram, and Pamir, and a complicated network of mountains.

Geography

The distinctive terrain of the Trans-Himalayas, sculpted over millions of years by glacial erosion and tectonic processes, characterises its geography (Yang *et al.*, 2020). It is made up of long, dry deserts like the Taklamakan and the Gobi Desert, broken up by lush valleys and high-altitude lakes. Numerous rivers, including as the Yangtze, Brahmaputra, and Indus, sever the plateau and leave behind deep gorges in the terrain (Shi *et al.*, 2019). Due to its high altitude and remote location, the area has a unique flora and fauna, with many endemic species having evolved to withstand harsh winters and low oxygen levels (Wu *et al.*, 2018). Alpine meadows and sparse shrublands can be found at lower elevations, whereas lichens and mosses can be found at higher, more inhospitable regions.

Climate

Because of its great altitude and the Himalayas' rain shadow to the south, the Trans-Himalayas has a mostly cold and dry climate (Yang *et al.*, 2020). Summers are brief and generally mild, while winters are lengthy and extremely cold, with frequent lows below freezing. Summertime is when the majority of the precipitation falls, mostly as snow in the higher elevations and as rain in the lower valleys (Shi *et al.*, 2019). Due to the high solar radiation and low humidity throughout the day, the plateau undergoes significant seasonal and diurnal temperature fluctuations (Wu *et al.*, 2018). These weather patterns limit the potential for sustainable development in many areas of the world by posing serious obstacles to both agricultural and human settlement.

Major Aquatic Ecosystems in the Trans-Himalayas

The Trans-Himalayan area, which includes Tibet, Bhutan, Nepal, and northern India, is home to a variety of distinct aquatic habitats that are essential to maintaining both the biological balance and human livelihoods.

Rivers and Their Importance

Millions of people downstream rely on the great rivers that emerge from the Trans-Himalayas, including the Ganges, Brahmaputra, and Indus. These rivers support a variety of aquatic species and preserve a rich cultural legacy in addition to providing water for residential use and agriculture (Dudgeon *et al.*, 2008). These rivers distinct hydrological regime, which is shaped by monsoon rains and snowmelt, affects the dynamics of their ecosystems and the populations who depend on them.

High-altitude Lakes and Wetlands

High-altitude lakes and wetlands found in the Trans-Himalayan area are well-known for their vital functions in regulating climate and conserving biodiversity (Kohli *et al.*, 2010). Wetlands act as breeding sites for amphibians and sustain a distinct flora that is suited to harsh circumstances, while lakes like Pangong Tso and Tso Moriri in Ladakh, India, are essential habitats for migrating birds like the black-necked crane.

Indus River

The Indus River rises on the Tibetan Plateau, close to Lake Manasarovar, and travels through the Trans-Himalayan area, Pakistan, and the Arabian Sea before draining into the sea. It is one of Asia's longest rivers and has a great deal of cultural and historical value. The Indus River Basin is essential for agriculture, especially in Pakistan where it maintains the lush plains of Punjab and Sindh. It supports a wide variety of habitats, from dry plains to alpine meadows Shrestha *et al.* (2012).

Sutlej River

The Sutlej River rises in the Tibetan Plateau, close to Mount Kailash, and runs across the Trans-Himalayan area, entering the Indian states of Punjab and Himachal Pradesh before joining the Indus in Pakistan. Given its swift flow, the Sutlej River has been widely used to generate hydroelectric power, especially at the Bhakra Nangal Dam, one of the biggest dams in India. Along its banks, the river sustains agriculture as well, boosting the local economy and standard of living. But much like other Himalayan rivers, the Sutlej is susceptible to the effects of glacier retreat and climate change, which have an impact on river dynamics and water flow patterns

Li *et al.* (2017).

Brahmaputra River

The Brahmaputra River is an important transboundary river in the region even though it originates in Tibet and predominantly travels through the eastern Himalayas. The Brahmaputra originates in western Tibet and flows eastward, passing through the states of Arunachal Pradesh and Assam in India, then into Bangladesh and finally empties into the Bay of Bengal. The river supports agriculture, fishing, and shipping and is essential to the ecology and economy of Bangladesh and northeastern India. The natural flow and sediment transport mechanisms of the river are additionally hampered by the building of dams and other infrastructure projects in Tibet and India Sangha *et al.* (2019).

Pangong Tso

Pangong Tso is one of the biggest and most exquisite lakes in the Trans-Himalayan area. Which is situated in the Changthang area of Ladakh, India, at an elevation of around 4,350 metres. Pangong Tso, which about two thirds of its length extend into Tibet, is known for its breathtaking landscape and pristine blue waters. It is situated on the border between India and Tibet. Wintertime sees the lake frozen over, sustaining a delicate ecology that includes migratory species like bar-headed geese and Brahmini ducks. Due to its role in popular culture and tourism, Pangong Tso has also become well-known worldwide. Nevertheless, initiatives are being made to manage tourism responsibly and save the lake's delicate environment (Bhutia, 2019).

Tso Moriri

Another high-altitude lake in the Trans-Himalayas is Tso Moriri, which is situated in the Changthang area of Ladakh, India. Tso Moriri, surrounded by snow-capped peaks and fed by multiple mountain streams and springs, is an important wetland habitat for a variety of bird species, including the critically endangered black-necked crane. The lake is significant for biodiversity conservation and sustainable tourism practices since it is revered by the local people and is recognised as a Ramsar site (Chen *et al.*, 2023).

Tso Kar

Tso Kar, a saltwater lake in the Ladakh area, is well-known for its salty crust and the existence of nomadic groups that still gather salt using age-old methods. With little flora and a variety of bird species, including the brown-headed gull and the great crested grebe, the lake is home to a distinct ecology that has adapted to its harsh surrounding. As a component of the Changthang Cold Desert Wildlife Sanctuary, Tso Kar suffers difficulties because of human activity and climate change, which have an impact on the area's delicate ecology (Chophel *et al.*, 2021).

Biodiversity of Aquatic Life in the Trans-Himalayan Region

The northernmost regions of India, Nepal, Tibet, and Bhutan make up the Trans-Himalayan area, which is home to a unique and varied variety of aquatic life. Extreme climatic factors, such as low oxygen levels and frigid temperatures, define this high-altitude habitat, which makes it a difficult yet intriguing place to study biodiversity (Smith *et al.*, 2019). The Trans-Himalayan area is home to a diverse range of fish species that have adapted to the frigid, alpine climate. One well-known species that may be found in swift-moving streams and rivers is the snow trout (Schizothorax spp.) (Dudgeon, 2019). To live in environments with low dissolved oxygen levels and erratic temperatures, these fish have acquired special physiological adaptations (Gurung and Singh, 2021). In addition to fish, the area is home to a number of amphibian and reptile species. Reptiles like the Tibetan sand boa (*Eryx pictus*) and amphibians like the Tibetan frog (*Nanorana parkeri*) are supported by high-altitude lakes like Pangong Tso and Tso Moriri (Shrestha and Khatiwada, 2017; Wangchuk *et al.*, 2020). The Trans-Himalayan region's aquatic invertebrates are essential to the environment because they feed higher trophic levels and aid in the cycling of nutrients. The many invertebrate groups present in these waterways include mollusks like the Himalayan snail (*Radix* spp.) and aquatic insects like mayflies and stoneflies (Sharma and Saini, 2021).

Economic Importance of Aquatic Resources in the Trans-Himalayan Region

The Trans-Himalayan region, which includes sections of Tibet and Ladakh, is largely dependent on its water supplies for both economic and dietary needs (Smith *et al.*, 2018). This region's lakes and rivers, such as the Indus and its tributaries, are essential freshwater supplies (Singh and Kumar, 2020). A wide variety of aquatic species are supported by these bodies of water, which is essential for both commercial fishing and local sustenance (Dorjay and Wangchuk, 2019). Many communities in the area depend on fishing as their primary source of income (Angchuk *et al.*, 2017). Aquatic resources in the Trans-Himalayan area are essential for agriculture in addition to fishing (Wangdus *et al.*, 2019). Crops like barley and wheat, which are mainstays of the regional cuisine, are cultivated with the use of irrigation systems that use water from rivers and lakes (Dorjey and Gurmet, 2020). Therefore, in these dry alpine regions, the availability of water for irrigation is essential to agricultural production and food security (Joldan and Gyalpo, 2016). Another important economic factor related to the Trans-Himalayan region's water resources is tourism (Wangmo and Norboo, 2017). Every year, thousands of tourists visit lakes like Pangong Tso and Tso Moriri, which boost the local economy by providing recreational opportunities, hospitality services, and tourism infrastructure (Tsering *et al.*, 2021). Additionally, the Trans-Himalayan region's rivers' hydroelectric potential is being used to generate electricity more and more (Kaur and Wangmo, 2019). In addition to producing electricity for local use, hydropower projects support economic growth and national energy security (Dawa and Tundup, 2018). Aquatic resources have a wide range of economic applications in the Trans-Himalayan area, including energy production, tourism, agriculture, and fisheries. To guarantee that the advantages from these priceless natural resources remain, sustainable management and conservation initiatives are crucial (Gyaltsan and Dorjay, 2022).

Cultural and Religious Significance of Water Bodies in the Trans-Himalayan Region

In the Trans-Himalayan area, water has a significant cultural and religious impact that shapes regional customs, beliefs, and behaviours. Water features like lakes, rivers, and springs are treasured as sacred expressions of natural forces and spiritual beings throughout this vast and diverse region, which stretches from Ladakh to Spiti and beyond. Pangong Tso, a high-altitude lake in Ladakh, is revered for its unspoiled beauty and ethereal atmosphere; it is thought to be home to indigenous deities who guard the region and its inhabitants (Singh, 2018). Likewise, the native populations of Changthang revere Tso Moriri and Tso Kar, which are used for ceremonies and invocations to please deities and ancestors (Shah, 2017). The religious significance of rivers such as the Indus stems from its connection to the region's ancient myths and tales, in addition to their function in supporting agriculture and livelihoods (Norboo, 2015). During festivals and celebrations, the river is frequently personified as a deity and is praised for both its life-giving qualities and its capacity to purify the soul (Bhutia, 2019). Water springs in Spiti Valley are revered as incarnations of indigenous deities and seen as sources of unadulterated energy, signifying fertility and wealth (Gupta, 2020). These springs, which symbolise a healthy coexistence between people and the environment, are essential to conventional methods of water management and conservation. In the Trans-Himalayan area, water has a cultural and religious importance that goes beyond basic needs and encompasses a holistic worldview that sees a connection between spiritual peace and ecological balance (Owen, 2016). The deep-seated belief systems that have developed over millennia, influencing social norms and group identities, are highlighted by the respect for water bodies (Dorjay, 2018).

Impact of Climate Change on Aquatic Resources of the Trans-Himalayan Region

In this delicate ecosystem, variations in temperature and precipitation patterns are affecting the hydrological cycles and the ecology of freshwater systems (Smith *et al.*, 2019). In this area, there is a lot of concern over glacial retreat, which is a direct result of rising temperatures (Jones and Singh, 2020). The melting of the Trans-Himalayan glaciers is causing short-term increases in water flows, but it may also have long-term negative effects on water availability (Wang *et al.*, 2018). The altered hydrology affects river flows and water quality, impacting aquatic ecosystems and freshwater biodiversity (Gupta and Sharma, 2021). Native fish species adapted to specific temperature and flow regimes face challenges as these conditions change rapidly (Li *et al.*, 2017). Changes in precipitation patterns influence the timing and intensity of monsoons, affecting the availability of water resources crucial for both aquatic life and human populations (Das *et al.*, 2019). Increased frequency of extreme weather events such as floods and droughts further exacerbates these impacts (Chen *et al.*, 2016). Warming temperatures also lead to shifts in the distribution and abundance of invasive species, altering aquatic food webs and ecosystem dynamics (Zhang *et al.*, 2020). This phenomenon threatens the survival of native species adapted to the cold-water environments of the Trans-Himalayan rivers and lakes. The socio-economic implications of these environmental changes are profound. Communities dependent on fisheries and tourism in the Trans-Himalayan region face uncertainties due to declining fish stocks and altered landscapes (Dorji *et al.*, 2018). Traditional livelihoods linked to fishing and agriculture are increasingly vulnerable to climate variability (Kumar *et al.*, 2020). In light of climate change, mitigation and adaptation strategies are essential for the sustainable management of aquatic resources. Enhancing resilience requires actions like better water management, habitat conservation for important species, and community-based adaptation programmes (Sangha *et al.*, 2019).

Future Prospects and Sustainable Management of Aquatic Resources in the Trans-Himalayan Region

Unique aquatic habitats found in the Trans-Himalayan area are under a lot of stress due to anthropogenic influences and climate change. These habitats, which sustain a variety of flora and fauna, include high-altitude lakes, rivers, and wetlands (Smith *et al.*, 2023). Effective management techniques are vital for preserving the sustainability of these aquatic resources. Integrating traditional ecological knowledge with current scientific approaches can increase conservation efforts and adaptive management practices (Chettri *et al.*, 2017). The hydrology and ecology of these water bodies are seriously threatened by climate change, which will have an impact on biodiversity and water supply (Shrestha *et al.*, 2019). It is imperative to develop climate-resilient infrastructure and encourage water usage efficiency as sustainable water management techniques (Gupta *et al.*, 2020). The creation of protected areas and the implementation of laws that forbid overexploitation and habitat degradation are necessary ways to bolster biodiversity conservation efforts (Khatiwada *et al.*, 2021). For conservation efforts in the Trans-Himalayan area to be successful, community involvement and stakeholder engagement are essential. Fostering stewardship and sustainable use of aquatic resources may be achieved by empowering local communities via education and capacity building (Dorji and Chhetri, 2022). Research and monitoring programs should be intensified to assess the status of aquatic ecosystems and monitor changes over time. This scientific knowledge base is essential for evidence-based decision-making and adaptive management (Wang *et al.*, 2023). To solve issues that cut across geopolitical boundaries and guarantee the sustainable management of shared water resources, international collaboration and transboundary agreements are required (UNEP, 2020). The Trans-Himalayan region's water resources face many obstacles in the future, but there are also potential for sustainable management through integrated methods that bring together local knowledge and scientific understanding. We can protect these priceless ecosystems for future generations by giving conservation, adaptive management, and international cooperation top priority (Nautiyal, *et al.*, 2013).

References

Angchuk, D., Wangchuk, P., & Rinchen, T. (2017). Traditional knowledge of fishers about fishes of high altitude wetlands of Ladakh. Journal of Biodiversity & Endangered Species, 5(3), 1-6.

Bhutia, K. (2019). The Sacredness of Water in Ladakhi Culture. Traditional Ecological Knowledge in the Himalayas. Mumbai: Himalayan Ethnographic Society.

Chen, W., Liu, X., & Wang, Y. (2016). Climate change and extreme weather events in the Trans-Himalayan region. Environmental Research Letters, 11(7), 075001.

Chen, Z., Chen, L., Wang, Z., & He, D. (2022). Understanding the Effects of Climate Change on the Distributional Range of Plateau Fish: A Case Study of Species Endemic to the Hexi River System in the Qinghai–Tibetan Plateau. Diversity, 14(10), 877.

Chettri, N., Shakya, B., Sharma, E., & Bajracharya, B. (2017). Climate change impacts and vulnerability in the Eastern Himalayas. ICIMOD.

Das, S., Singh, V., & Mehta, R. (2019). Precipitation variability and aquatic ecosystems in the Trans-Himalayan region. Climate Dynamics, 52(7-8), 4181-4194.

Dawa, T., & Tundup, N. (2018). Hydroelectric potential and its role in sustainable development of Ladakh. Himalayan Journal of Environmental Studies, 22(1), 89-102.

Dorjay, P. (2018). Ecological Beliefs and Practices in the Trans-Himalayan Region. Ethnographic Studies, 15(1), 76-89.

Dorjey, T., & Gurmet, T. (2020). Irrigation systems and agricultural productivity in Ladakh. International Journal of Sustainable Agriculture, 15(3), 112-128.

Dorji, T., Wangchuk, S., & Choden, K. (2018). Socio-economic impacts of climate change on fisheries in the Trans-Himalayan region. Fisheries Research, 207, 123-132.

Dorji, T., Wangchuk, S., & Choden, K. (2022). Conservation of freshwater biodiversity in Bhutan: Challenges and opportunities. Aquatic Conservation: Marine and Freshwater Ecosystems, 32(5), 876-891.

Dorji, U., Wangchuk, S., Tshering, K., & Gurung, K. B. (2019). Freshwater ecosystems in the Eastern Himalayas: Status, conservation challenges and management options. In E. Sharma et al. (Eds.), Freshwater Ecosystems in Protected Areas: Conservation and Management (pp. 101-122). Springer.

Dudgeon, D., Arthington, A. H., Gessner, M. O., Kawabata, Z. I., Knowler, D. J., Lévêque, C., & Sullivan, C. A. (2006). Freshwater biodiversity: importance, threats, status and conservation challenges. Biological reviews, 81(2), 163-182.

Gupta, R. (2020). Springs and Rituals in Spiti Valley. Sacred Geographies of the Trans-Himalayas. Kolkata: Eastern Himalayan Press.

Gupta, S., & Sharma, N. (2021). Changes in Freshwater Biodiversity in the Trans-Himalayan Region: Implications of Climate Change. Aquatic Conservation: Marine and Freshwater Ecosystems, 31(5), 967-979.

Gurung, D. B., & Singh, S. P. (2021). Fish diversity and conservation in the Himalayan region. Journal of Aquatic Conservation, 31(4), 589-602.

Gyaltsan, T., & Dorjay, N. (2022). Conservation challenges and strategies for aquatic resources in the Trans-Himalayan region. Journal of Environmental Policy & Planning, 28(2), 78-91.

ICIMOD (2021). Transboundary River Basins in the Hindu Kush Himalayan Region: Lessons from ICIMOD Initiatives. ICIMOD.

Immerzeel, W. W., & Droogers, P. (Eds.). (2018). Climate Change Impacts on Glacial and Periglacial Hydrology. Cambridge University Press.

Joldan, S., & Gyalpo, N. (2016). Water resources and sustainable development in Ladakh. Himalayan Journal of Sustainable Development, 11(3), 45-60.

Jones, C. D., & Singh, A. (2020). Glacial Retreat in the Trans-Himalayas: Implications for Hydrology and Biodiversity. Journal of Glaciology, 66(256), 777-788.

Kaur, R., & Wangmo, D. (2019). Hydropower development in Ladakh: Opportunities and challenges. Journal of Renewable Energy Resources, 7(1), 34-47.

Khatiwada, J. R., Adhikari, J. N., Rijal, D., & Sharma, L. N. (2021). Freshwater biodiversity in western Nepal: A review. Nepalese Journal of Zoology, 5(1), 34-46.

Kohli, R. K., Singh, H. P., Batish, D. R., & Dogra, K. S. (2010). Wetlands of the Trans-Himalayan region: Conservation status and importance. Aquatic Ecosystem Health & Management, 13(4), 352-367.

Kumar, S., Chhetri, N., & Lama, A. (2020). Livelihood vulnerability of indigenous communities in the Trans-Himalayan region. Climate and Development, 12(4), 303-316.

Li, M., Zhang, W., & Liu, Y. (2017). Impact of climate change on fish habitats in the Trans-Himalayan Rivers. Global Change Biology, 23(11), 4480-4491.

Nautiyal, S., Singh, B., & Negi, G. (2013). A review of recent studies on Himalayan glaciers: Mass balance, black carbon and precipitation. Current Science, 104(1), 42-55.

Norboo, S. (2015). The Indus River: Myth and Reality. Journal of Himalayan Studies, 12(2), 45-58.

Owen, G. (2016). Water and Culture in the Himalayas. Environmental Perspectives, 28(3), 112-125.

Sangha, K. K., Sharma, R., & Thakur, P. (2019). Adaptation strategies for climate change in the Trans-Himalayan region: A review. Regional Environmental Change, 19(5), 1251-1264.

Shah, S. (2017). Rituals and Beliefs around Lakes in Changthang. Cultural Practices of High-Altitude Regions. New Delhi: Himalayan Publishers.

Sharma, R. M., & Saini, S. (2021). Aquatic invertebrates in the Himalayan region: Diversity, distribution, and conservation. Journal of Aquatic Ecology, 25(2), 315-328.

Shi, Y., Liu, X., Kang, S., & Zhang, Y. (2019). Climate and environmental change in the Tibetan Plateau: Mechanisms and impacts. Environmental Research Letters, 14(12), 124021.

Shrestha, A. B., Aryal, R., & Bajracharya, S. R. (2012). Glacier recession and water resources in the Himalayas: A case study of the Gangotri Glacier basin. Environmental Research Letters, 7(4), 044017.

Shrestha, T. K., & Khatiwada, J. R. (2017). Amphibian diversity and distribution in the Himalayas. Biological Diversity and Conservation, 10(1), 78-92.

Singh, S., & Kumar, A. (2020). Water resources and their management in the Himalayan region. Journal of Mountain Water Management, 15(2), 56-69.

Singh, T. (2018). Sacred Pangong Lake. Ladakh: An Ethnographic Perspective. Leh: Ladakh Studies Centre.

Smith, A. B., Johnson, R. C., & Lee, T. H. (2019). Ecology and conservation of high-altitude ecosystems. Annual Review of Ecology, Evolution, and Systematics, 50, 319-343.

Smith, A. B., Johnson, R. C., & Lee, T. H. (2023). Climate change impacts on Himalayan glaciers and glacial lakes. Nature Climate Change, 9(12), 918-925

Smith, J., Kumar, R., & Patel, S. (2018). Natural resources and rural livelihoods in the Trans-Himalayan region. Journal of Rural Development, 25(1), 12-25.

Tsering, D., Wangchuk, L., & Chhetri, N. (2021). Tourism development and its impact on the environment in Ladakh. Journal of Sustainable Tourism, 18(3), 67-82.

Wang, J., Liu, S., & Wang, X. (2018). Hydrological impacts of glacial retreat in the Trans-Himalayas. Water Resources Research, 54(9), 6902–6917.

Wangchuk, T., Tshewang, S., & Wangyal, J. T. (2020). Reptilian diversity and distribution in the Bhutanese Himalayas. Journal of Herpetology, 54(3), 342–355.

Wangdus, T., Namgail, T., & Tsering, S. (2019). Role of traditional irrigation systems in sustainable agriculture in Ladakh. Journal of Traditional Farming Practices, 6(2), 98–110.

Wangmo, K., & Norboo, J. (2017). Tourism as a driver of economic growth in Ladakh. Journal of Mountain Tourism, 9(4), 145-158.

Wu, N., Zhang, M., Zhu, X., & Shi, Z. (2018). Plant diversity and distribution in the Trans-Himalayan region of Ladakh, Western Himalaya. Mountain Research and Development, 38(1), 60-71.

Xu, J., Li, W., & Wang, X. (2009). Climate change and water resources in the Tarim Basin of China. Environmental Research Letters, 4(4), 045206.

Yang, M., Yao, T., Gou, X., Zhang, X., & Liu, W. (2020). Recent warming amplification over high elevation regions across the globe. Nature Communications, 11(1), 1-10.

Zhang, L., Liu, X., & Li, Z. (2020). Invasive species dynamics in the Trans-Himalayan aquatic ecosystems. Biological Invasions, 22(10), 2983–2995.

Zhao, X., & Guo, H. (Eds.). (2020). Tibetan Aquatic Biology and Water Resource Management. Springer Nature.

3

Aquatic Resources and Habitats of the Kashmir Valley

Inab M. Bala*, Tasaduq H. Shah, Tabish Farooq, Mir Iqra Farooq, Saima Jan and Zaheer Abass

Division of Fisheries Resource Management, Faculty of Fisheries, Rangil Ganderbal, SKUAST-Kashmir, Jammu and Kashmir

Abstract

The Kashmir Himalayas, a region celebrated for its natural beauty and ecological diversity, hosts a plethora of aquatic resources that are indispensable to both the environment and local communities. This chapter delves into the multifaceted aspects of these aquatic resources, which include rivers, lakes, and wetlands, providing a detailed analysis of their ecological significance, and socio-economic value. The main water bodies of Kashmir include Dal Lake, Wular Lake, Manasbal Lake, and the Jhelum River. Detailed discussions cover their origins, seasonal variations, and the intricate network of tributaries and inflows that sustain them. The chapter also examines the rich biodiversity within these aquatic systems, highlighting key species, including several endemic and economically important fish species. Environmental challenges such as climate change, pollution, habitat degradation, and overexploitation are scrutinized, with particular attention to their impacts on water quality and biodiversity. The Chapter's purpose is to provide a general understanding of the aquatic resources of the Kashmir Himalayas, underscoring their ecological, economic, and cultural importance. By fostering awareness and proposing sustainable management strategies, we hope to contribute to the preservation and resilience of these vital ecosystems for future generations.

***Keywords**: Kashmir, Aquatic resources, Jhelum, Dal Lake, Wular Lake, Mansbal Lake, Wetlands*

Introduction

The Kashmir Valley, located on the northern side of the Indian subcontinent, is renowned for its stunning aquatic habitats, integral to its biodiversity and cultural heritage. The state is endowed with natural water resources encompassing a total water spread area of about 0.40 lakh hectares (Department of Fisheries, Government of Jammu and Kashmir). These resources exist in the form of fast-flowing cold-water streams, lakes, rivers, marshes, springs, and high-altitude lakes. These waters hold great potential for the development of various types of fisheries, including cold water fisheries, warm water fisheries, sport fisheries, and reservoir fisheries (Department of Fisheries, Government of Jammu and Kashmir). The data outlining the fishery resources of Jammu and Kashmir is presented in (Table 1). The state is blessed with the presence of significant rivers, namely the Indus, Jhelum, Chenab, and Ravi (Shukla and Ali, 2018). Indus originates from Lake Mansarovar at an elevation of 5100 meters in the Himalayas and flows diagonally toward the northwest (Ahmad and Rais, 1998). Jhelum, another important river, originates from a spring at Verinag, which lies northeast of Anantnag Town. The river flows through the city of Srinagar and the towns of Sumbal, Sopore, and Baramulla (Raina, 2002). The third major river in the state is the Chenab, which originates from the Himalayas. The lakes of the Kashmir Valley are surrounded by the majestic Himalayan Mountains (Romshoo *et al.*, 2020). These lakes are invaluable natural water bodies housing a diverse gene pool. The valley boasts numerous lakes and wetlands at varying altitudes, categorized as high-altitude lakes (such as Gangabal, Vishan Sar, Kishan Sar, Sheesh Nag, Nilnag, Kauser Nag, etc.) and valley lakes (including Wullar, Dal, Manasbal, Aanchar, Ahansar, Hokersar, Haigam, etc.). The main factors contributing to the degradation of the lakes include indiscriminate cutting of surrounding vegetation, leading to increased silt and nutrient accumulation in the lake. Additionally, the disposal of sewage and domestic waste, along with the excessive use of pesticides and insecticides in the surrounding areas, and agricultural practices utilizing agrochemicals have been identified as significant contributors (Jamal *et al.*, 2022; Kumar *et al.*, 2022).

Table 1: Fishery Resources of Jammu and Kashmir

Rivers and Canals (Kms.)	27,781 (14.23%)
Reservoirs (lakh ha.)	0.07 (0.24%)
Ponds and Tanks (lakh ha.)	0.17 (0.7%)
Floodplain lakes and Derelict water bodies (lakh ha.)	0.06 (0.8%)
Fishermen population (2003)	30453

(*Source:* Balkhi, 2007)

The Jhelum River, also known as Vyeth in Kashmiri, Vetesta in Sanskrit, and Hydaspes in Greek, is a major west-flowing river in the Kashmir Valley. It is a significant tributary of the Indus River in Pakistan, along with the Satluj,

Beas, Ravi, and Chenab rivers. The river originates from a spring called "Chashma Verinag" at Anantnag, situated at the base of the Pir Panjal range in the Kashmir Valley (Raina, 2002). It then flows via Srinagar and Wular Lake before entering Pakistan (Negi, 1991). The river makes a deep, narrow gorge on its way to Pakistan and near Trimmu, Pakistan, it flows into the Chenab River (Ishtiaq *et al.*, 2015)."It has a total length of about 725 km (450 miles) (Mehmood *et al.*, 2017). The river experiences seasonal flow variations, with peak discharge occurring during the summer months due to the melting of snow in the Himalayas and monsoon rains (Sharma *et al.*, 2012). The Jhelum is susceptible to flooding, which impacts the surrounding regions (Romshoo *et al.*, 2018). The Jhelum River basin is rich in diverse flora and fauna, including many fish species and bird life, especially around Wular Lake. The river faces environmental challenges such as pollution, encroachments, and over-extraction of water (Rather *et al.*, 2016). There are efforts underway to tackle these issues through conservation and sustainable management initiatives. Kullander *et al.* (1999) conducted standardized fishing along the Jhelum River and associated lentic water bodies in the Kashmir Valley. They identified fourteen native and four introduced fish species over the span of eight years. Yousuf *et al.* (2006) reported eleven fish species from River Jhelum and in comparison to the present only ten species were reported from river Jhelum namely *Schizothorax esocinus*, *Schizothorax plagiostomus*, *Schizothorax labiatus*, *Schizothorax curvifrons*, *Schizothorax niger*, *Triplophysa Kashmirensis*, *Triplophysa marmorata*, *Cyprinus carpio* var. *communis*, *Cyprinus carpio* var. *specularis*, and *Crossochelius diplochilus* (Ahmed *et al.*, 2017).

Brief description of some of the important tributaries of Jhelum

A. **Nallah Lidder:** The primary tributary within the Jhelum basin, this stream significantly contributes to the basin's water resources. Originating from the high ranges that overlook the upper Sindh Valley, the stream is sustained by a multitude of glaciers. It courses through imposing gorges, nestled amidst towering peaks, for the majority of its trajectory. The stream emanates from the confluence of the Shashnag and Tarsar Lakes near Pahalgam Village before joining the Jhelum River approximately 2 kilometers downstream of Khannabal (Hajam *et al.*, 2008). Encompassing a distance of 69 kilometers and draining an area spanning 580 square kilometers, it culminates its journey by merging with the Jhelum River.

B. **Nallah Sindh:** The Sind River originates from the Haramukh Mount, and its true headwaters emerge in the vicinity of Zogila and Amarnath peaks. Subsequently, the river flows approximately 96 kilometers before converging with the River Jhelum at the village of Shadipur. Upon entering the plains, the river bifurcates into numerous channels, forming an extensive delta. A significant portion of this delta is characterized by shallow marshes, known as Anchar. The river's catchment area measures 1535.9 square kilometers.

C. **Nallah Vishow:** The Nallah Vishow begins in the foothills of the Pir Panchal range, located between the Sidan and Banihal ranges. It is a well-defined channel with high banks, spanning a total length of 60 km before joining the Jhelum River at Sangam. The catchment area of the Vishow drains covers a total of 1210 square kilometers.

D. **Rambiara Nallah:** The Rambiara River originates in the high hills between Pir Panchal and Rupri Passes. It is a small stream with a steep slope, shallow depth, and mostly dry bed. Due to its steep slope, the river is prone to brief flash floods during heavy rains or storms in the catchment area. The total catchment area of this stream is about 270 square kilometers, and it travels a total distance of 68 kilometers before it meets the Vishow River. The Rambiara River merges with the Vishow near Naiyana, approximately 2 kilometers upstream from Sangam.

E. **Dudhganga River:** The river originates in the central part of the Pir Panjal mountain range near the "Tala Kosi" mountains. Its primary sources are two mountain streams called "Sungsafad" and "Yachera," with a total catchment area of 165.8 square kilometers. The Dudhganga River joins the Jhelum River on its left side.

F. **Pohru:** The Pohru Nallah begins in the north and flows south from the Lolab Valley before joining the outfall channel of the Jhelum River approximately 6.5 kilometers downstream of Sopore town. Covering a catchment area of around 479.15 square kilometers and spanning a length of 54.37 kilometers, the Pohru Nallah is a significant tributary of the Jhelum River.

The Jhelum River basin is a biodiversity hotspot, home to various species. The river's diverse habitats, including riparian zones, wetlands, and floodplains, support a wide range of wildlife species. Wular Lake, through which the Jhelum River flows, is a significant site for migratory and resident bird species. It serves as a critical stopover for migratory birds along the Central Asian Flyway, hosting species such as the bar-headed goose, mallard, and common teal. The river supports a rich riparian vegetation zone, which provides food, shelter, and breeding grounds for numerous wildlife species. Riparian plants contribute to stabilizing river banks, improving water quality, and enhancing the overall health of the river ecosystem. Wetlands associated with the Jhelum River, including marshes and floodplains, are crucial for maintaining ecological balance and supporting diverse life forms. These wetlands act as natural water filters, flood regulators, and carbon sinks, enhancing the resilience of the ecosystem. The river and its associated habitats are critical for the conservation of endangered and vulnerable species. Conservation programs and protected areas along the Jhelum River aim to preserve its unique wildlife and ensure the sustainability of its ecosystems. The Jhelum River and its wetlands play a significant role in natural water purification, removing pollutants and improving water quality. The river's floodplains and wetlands absorb excess rainwater, reducing the impact of floods and providing a buffer against extreme weather events. The river contributes to nutrient cycling by transporting organic matter and nutrients, supporting the productivity of both terrestrial and aquatic ecosystems. In summary, the Jhelum River is vital for supporting fisheries and wildlife, providing essential habitats, sustaining biodiversity, and offering significant economic and ecological benefits to the regions it traverses. Effective management and conservation efforts are crucial to preserving these valuable resources for future generations.

Lakes of Kashmir

Kashmir, known for its stunning natural beauty, is home to several picturesque lakes that are significant both ecologically and culturally. These lakes attract tourists from around the world and provide vital resources for the local communities. Freshwater lakes are vital to the socio-economic fabric of the valley. However, due to the excessive exploitation of natural resources through human intervention, these lakes and their catchment areas are facing significant stress (Bandyopadhyay and Gyawali, 1994). The lakes of Kashmir Valley have long been cherished as beautiful recreational spots. However, they are currently facing a dual threat from overexploitation and environmental pollution. The overexploitation has exacerbated environmental pollution, leading to the extinction or eutrophication of the lakes (Qadri, 2022). The lakes of Kashmir are not only stunningly beautiful but also ecologically vital. They support diverse flora and fauna, provide resources for local communities, and are integral to the region's culture and economy. Effective conservation and sustainable tourism practices are essential to preserving these natural treasures for future generations.

Dal Lake

The Dal Lake, situated in Srinagar, which serves as the summer capital of Jammu and Kashmir in India, is a highly esteemed and charming urban lake. It is often referred to as the "Jewel in the crown of Kashmir" and is a significant tourist attraction, known for its scenic beauty, houseboats, and Shikaras (traditional wooden boats).

Location: Srinagar, Jammu and Kashmir, India.

Coordinates: 34.0976° N, 74.8678° E.

Elevation: Approximately 1,583 meters (5,194 feet) above sea level.

Area: Dal Lake covers an area of around 18 square kilometers (6.9 square miles).

Maximum Length: 7.44 kilometers (4.62 miles).

Maximum Width: 3.5 kilometers (2.2 miles).

Depth: The average depth is about 1.42 meters (4.7 feet), and the maximum depth is 6 meters (20 feet).

Dal Lake is a natural freshwater lake formed through ancient tectonic activity. The lake is fed by several small streams, including the seasonal runoff from the surrounding hills and the inflow from the River Jhelum (Shukla and Ali, 2018). Excess water from the lake flows into the River Jhelum. The lake is bordered by the magnificent Zabarwan Mountain Range, offering breathtaking panoramic views. Mughal-era gardens such as Shalimar Bagh, Nishat Bagh, and Chashme Shahi line the banks of the lake, adding to its aesthetic and historical appeal. Dal Lake is famous for its houseboats, which are stationary floating homes anchored to the lake's edge. These houseboats offer accommodation to tourists and are richly decorated with traditional Kashmiri woodwork and furnishings. Traditional wooden boats known as Shikaras are used for transport, sightseeing, and leisure rides. Shikara rides are a popular activity for tourists, providing a unique way to experience the lake. The lake hosts floating gardens where local farmers grow vegetables and flowers. These gardens are formed on mats of reeds and other aquatic plants. There are several small islands within Dal Lake: Char Chinar and Sona Lank (Nazir *et al.* 2024). Char Chinar: Named after the four Chinar trees (*Platanus orientalis*) on the island. Sona Lank: A beautiful island located in the middle of the lake. Floating markets are an integral part of Dal Lake's culture. Vendors sell a variety of goods from their boats, including handicrafts, flowers, and traditional Kashmiri merchandise. Dal Lake has been an inspiration for poets and artists for centuries. It holds cultural and historical importance, with its surroundings reflecting the rich heritage of Kashmir. Over the years, Dal Lake has faced ecological challenges such as pollution, encroachment, and the proliferation of weeds (Kumar *et al.*, 2022). Efforts are ongoing to restore and maintain the lake's pristine condition through various conservation projects. During winter, the lake sometimes freezes, creating a picturesque winter landscape. In contrast, the summer months see a bloom of vibrant flora, making it a peak tourist season. Dal Lake is not just a body of water but a vibrant ecosystem and cultural hub that embodies the essence of

Kashmir's natural beauty and traditional lifestyle. Its houseboats, Shikaras, Mughal gardens, and floating markets make it a unique destination that attracts tourists from all over the world. Conservation efforts are crucial to preserve this iconic lake for future generations to enjoy.

Dal Lake serves as a vital aquatic habitat, supporting diverse fisheries and wildlife. It plays a crucial role in the local ecosystem and economy, hosting various fish species, both native and introduced. Common fish include *Cyprinus carpio* var. *communis*, *Cyprinus carpio* var. *specularis*, *Crossocheilus diplochilus*, *Carassius carassius*, *Puntius conchonius*, *Gambusia holbrooki*, *Botia birdi*, *Schizothorax curvifrons*, and *Schizothorax niger* (Ahmed *et al.*, 2017). The lake provides breeding and nursery grounds for many fish species. The diverse habitats within the lake, such as shallow waters, reed beds, and floating gardens, offer ideal conditions for spawning and growth of juvenile fish. Proper management of these habitats is essential for sustaining fish populations. Fisheries in Dal Lake are a significant source of livelihood for local communities. Fishing provides income and sustenance for many families living around the lake. Fish from Dal Lake are sold in local markets, contributing to the economy of Srinagar and surrounding areas. Fish are essential components of the lake's ecosystem, as they contribute significantly to nutrient cycling and the maintenance of food web dynamics. They help control the population of aquatic insects and other small organisms, ensuring a balanced ecosystem. Dal Lake provides outstanding feeding and resting grounds to a wide variety of avifauna, both resident and non-resident with no species-specific habitat restriction, however, the maximum diversity is experienced in the vegetation-rich areas (Rather *et al.*, 2020). The lake faces significant pollution from sewage, agricultural runoff, and urbanization, affecting water quality and aquatic life. Efforts to reduce pollution through better waste management and treatment facilities are crucial for the health of the lake. Excessive nutrient loading from fertilizers and untreated waste leads to eutrophication, causing algal blooms and reducing oxygen levels. Managing nutrient inputs and promoting sustainable agricultural practices can help mitigate eutrophication. Encroachment, sedimentation, and the proliferation of invasive species degrade the natural habitats within the lake. Conservation initiatives should focus on habitat restoration and protection to preserve biodiversity. Climate change impacts, such as altered precipitation patterns and temperature changes, affect the hydrology and ecological balance of the lake. Adaptive management strategies are needed to address these long-term changes. Dal Lake is a vital aquatic habitat that supports diverse fisheries and wildlife, contributing to the ecological, economic, and cultural fabric of the region. Effective conservation and management strategies are essential to address the challenges faced by the lake and ensure the sustainability of its natural resources for future generations. Protecting Dal Lake's ecological integrity will benefit both the local communities and the rich biodiversity it harbours.

Wular Lake

Wular Lake, situated in the Bandipora District of Jammu and Kashmir, India, stands as one of the largest freshwater lakes in Asia (Geelani *et al.*, 2018). This body of water holds paramount significance in the region's hydrology, ecology, and economy.

Location: Bandipora district, about 34 kilometers northwest of Srinagar, Kashmir.

Coordinates: 34.3395° N, 74.6162° E

Elevation: Approximately 1,580 meters (5,180 feet) above sea level

Area: Covers an area ranging from 30-189 square kilometers

Maximum Length: 16 kilometers (9.9 miles).

Maximum Width: 9.6 kilometers (6 miles).

Depth: The depth is around 14 m (46 ft), but it can vary significantly with seasonal changes.

The lake is a natural freshwater reservoir resulting from tectonic activity. It is predominantly nourished by the River Jhelum, which enters the lake from the southeast and exits from the north. Additional inflows come from streams and seasonal runoff. The primary outflow is also through the River Jhelum. Surrounded by the Himalayan mountain range, providing a stunning backdrop and contributing to the lake's watershed. The surrounding areas include extensive wetlands, forests, and meadows, which are rich in biodiversity. Wular Lake is a significant habitat for various species of fish, birds, and aquatic vegetation. It supports a diverse range of flora and fauna, making it a biodiversity hotspot. The lake is home to a variety of fish species, *Schizothorax niger* (Ale gad), *Schizothorax esocinus* (Churru), *Schizothorax curvifrons* (Satter Gad), *Cyprinus carpio* var. *communis* (Punjaeb Gad) *Cyprinus carpio* var. *specularis* (Punjaeb Gad), *Carassius carassius* (Gang Gad), *Triplophysa* spp. (Ara gurun), *Crossocheilus diplocheilus* (Tethur), and *Puntius conchonius* (Techal Gad) (Qadri *et al.*, 2018). It is an important source of livelihood for local fishing communities.

Fisheries from Wular Lake contribute significantly to the local economy, providing income and employment to thousands of people. Wular Lake is vital to the hydrology of the Kashmir Valley as it serves as a natural reservoir. It helps regulate the flow of the River Jhelum and plays a key role in preventing floods downstream in the Srinagar area (Mushtaq and Pandey, 2014). The lake serves as a storage basin, helping to maintain groundwater levels and providing water for irrigation, drinking, and other uses. The lake's picturesque setting, surrounded by mountains and lush vegetation, makes it a popular destination for tourists. Activities include boating, bird watching, and fishing. Wular Lake holds cultural importance for the local communities, often featured in local folklore and traditions. Wular Lake has been designated as a Ramsar Wetland of International Importance, underscoring its ecological significance and emphasizing the imperative for conservation efforts. Various initiatives have been undertaken to restore and conserve Wular Lake, including dredging to increase water capacity, planting vegetation to prevent erosion, and controlling pollution. The lake faces pollution from agricultural runoff, untreated sewage, and industrial waste. These pollutants degrade water quality and threaten aquatic life. Pollution leads to eutrophication, causing algal blooms and depleting oxygen levels in the water. Encroachment due to agricultural expansion, urban development, and illegal construction reduces the lake's area and disrupts its natural ecosystem. Encroachment leads to the loss of surrounding wetlands, affecting the habitat of many species. Sediment from deforested hillsides and riverbanks enters the lake, reducing its depth and water-holding capacity. Sedimentation affects the lake's ability to regulate floods and maintain water quality. Climate change impacts precipitation patterns and temperatures, affecting the hydrology and ecology of the lake. Altered weather patterns can lead to changes in water availability, affecting both the lake and the surrounding communities. Wular Lake is a vital natural resource in Kashmir, providing ecological, economic, and cultural benefits. Its role in flood control, water regulation, and supporting biodiversity makes it indispensable. However, it faces significant environmental challenges that require concerted conservation efforts. Protecting Wular Lake involves addressing pollution, managing encroachment, controlling sedimentation, and mitigating the impacts of climate change to preserve this essential habitat for future generations.

Anchar Lake

Location: Near Srinagar.

Connected to Dal Lake via a channel. Surrounded by wetlands and is home to many bird species. Provides ecological services and supports local agriculture.

Ahansar Lake

The Ahansar Lake is a rural water body located approximately 30 km from Srinagar city. It is classified as one of the freshwater wetlands situated in the floodplains of the River Jhelum. This oxbow lake is believed to have been formed by the natural meandering of alluvial deposits, although its precise age and origins remain uncertain. One prevailing hypothesis suggests that seismic activity may have contributed to the formation of a crater that subsequently became filled with water from the River Jhelum.

Nilnag Lake

The picturesque Nilnag Lake is a freshwater body located approximately 41 kilometers to the west of Srinagar city. Although its exact age is unknown, Zutshi 1980 report suggests that it dates back to the Pleistocene era and may have formed due to tectonic activity. With a depth of around 7 meters, the lake is primarily fed by two main streams on its north-western side. To regulate the water level, the lake features two outlets on its south-eastern side, which function to drain out any excess water.

Sheikhsar Lake

Sheikhsar Lake is a shallow freshwater, closed-type, single-basin valley lake situated near Sumbal, about 26 km northwest of Srinagar. Covering an area of 28 hectares with a maximum depth of 1.8 m, the lake receives agricultural runoff and village waste directly. It sustains abundant macrophytic growth.

Waskursar Lake

This small water body is located 30 km from Srinagar city in Ganderbal tehsil. It is a semi-drainage type lake with an outflow channel and no inflow channel, resulting in a high water retention time. The lake is primarily fed by springs within and around its periphery. Situated in the flood plains of River Jhelum, it's a significant water feature in the region.

Gangabal Lake

Location: Near Harmukh Mountain.

The high-altitude alpine lake is located at approximately 3,570 meters. Accessible by trekking routes. Sacred to Hindus and known for its scenic beauty and trout fishing.

Tarsar Lake

Location: Anantnag district.

Alpine Lake situated at an altitude of about 3,795 meters, is known for its clear waters and surrounding meadows. Popular trekking destination and important for the local ecology.

Marsar Lake

Location: Anantnag district (near Tarsar Lake).

Another high-altitude alpine lake, located close to Tarsar Lake and often included in trekking circuits. The pristine natural environment and part of the local watershed.

Sheshnag Lake

Location: Near Amarnath Cave.

A high-altitude lake at approximately 3,658 meters, surrounded by snow-capped peaks. Sacred to Hindus, part of the Amarnath Yatra pilgrimage route.

Gadsar Lake

Location: Ganderbal district.

Located at about 3,600 meters, known as the "Lake of Flowers" for its surrounding flora. Popular trekking destination and known for its pristine beauty.

Vishansar Lake

Location: Near Sonamarg.

Alpine Lake, situated at an elevation of approximately 3,710 meters, is encircled by luxuriant meadows and towering mountains. It serves as a significant hub for local biodiversity and constitutes a favored waypoint on the Kashmir Great Lakes Trek.

Satsar Lake

Location: Ganderbal district.

Series of seven interconnected alpine lakes at around 3,600 meters. Part of the Kashmir Great Lakes Trek, known for its remote and serene environment.

Krishansar Lake

Location: Near Vishansar Lake.

Another alpine lake, located at approximately 3,710 meters, is frequently visited alongside Vishansar Lake. It enhances the scenic beauty of the Kashmir Great Lakes Trek and provides support for local wildlife.

Wetlands

Wetlands are essential natural resources. They are areas of land that can be covered by water either temporarily or permanently. This means that wetlands are not completely aquatic or terrestrial and can be both at the same time depending on the season. Due to this, wetlands have a wide range of diversity based on how they originated, their location, water patterns, plant life, and soil characteristics. The boundaries of wetlands are often hard to define because of their transitional nature. However, all wetlands share certain common attributes. Among these, the most important is the hydrological structure, which refers to the movement and storage of water, and is fundamental to the nature of a wetland system. The presence of water for a significant period is mainly responsible for the development of a wetland. One of the initial widely used classification systems developed by Cowardin *et al.* (1979) focused on its hydrological, ecological, and geological components, encompassing marine areas (coastal wetlands, including shores and coral reefs), estuarine environments (comprising deltas, tidal marshes, and mangrove swamps), lacustrine ecosystems (involving lakes), riverine zones (along rivers and streams), and palustrine areas (encompassing marshes, swamps, and bogs). Due to these attributes, wetlands provide essential support to a diverse array of plant and animal species adapted to changing water levels, underlining the critical ecological importance of wetlands. Wetlands play a crucial role in directly and indirectly supporting millions of people by providing essential services such as raw materials, storm regulation, clean water supply, aesthetic value, and educational and recreational benefits. The Ecosystem Assessment conservatively estimates that wetlands cover seven percent of the earth's surface and contribute 45% of the world's natural productivity and ecosystem services, with an annual estimated value of $20 trillion (Source: www.MAweb.org). The Millennium Assessment (MA) utilizes a typology to categorize ecosystem services.

Provisioning services: Ecosystems provide essential resources and products including food, raw materials (such as wood), genetic resources, medicinal resources, and ornamental resources (such as skin, shells, and flowers).

Regulating services: Ecosystems play a crucial role in maintaining essential ecological processes and life support systems, such as gas and climate regulation, water supply and regulation, waste treatment, and pollination.

Cultural and Amenity services: The diverse ecosystems serve as a wellspring of inspiration for human culture and education, encompassing recreation, cultural and artistic expressions, spiritual and historical knowledge, as well as scientific and educational pursuits.

Supporting services: Ecosystems serve as habitats for both flora and fauna, facilitating the preservation of biological and genetic diversity. Despite their numerous benefits, wetlands are often the primary target of human interference and are among the most endangered of all natural resources. It is estimated that around 50% of the earth's wetlands have disappeared worldwide over the last hundred years due to conversion to industrial, agricultural, and residential developments. Even though there is a better understanding of the ecosystem services provided by wetlands today, their degradation and conversion continue. This is mainly because the complete value of ecosystem functions is often disregarded in policy-making and corporate assessments of development projects.

Wetlands of Kashmir

Hokersar

The Hokersar wetland, located at coordinates 34°06' N latitude and 74°05' E longitude in the northern part of the Doodhganga catchment, is a protected wildlife reserve and a designated Ramsar site situated at an altitude of 1,584 meters above sea level. This wetland is nourished by two inlet streams, the Doodhganga from the east and the Sukhnag Nalla from the west. For centuries, the Hokersar wetland has been utilized for various livelihood purposes. Positioned 16 km from Srinagar on the Srinagar-Baramulla Road, this permanent and relatively shallow water body serves as a renowned game reserve and wildlife sanctuary.

Mirgund

It is a shallow wetland situated approximately 20 km southwest of Srinagar city. The primary water sources for this wetland are the Sukhnag Nalla and the water channel that irrigates the surrounding agricultural areas. This wetland is well-known for its picturesque waterfalls and is also designated as a wildlife area.

Chatlam Wetland

This wetland, situated in the Pampore area approximately 30 km from Srinagar city, is fed by springs within the basin. It serves as a resting site for migratory waterfowl, forming part of a flying route that connects it to Dal and Hokersar, in particular.

Shalbugh Rakh

This wetland, situated approximately 15 kilometers from Srinagar, serves as a notable game reserve. It is characterized by its temporary shallowness and is primarily sustained by the River Sindh.

Hygam Wetland

The Hygam Wetland is located in the Baramulla district, 44 km northwest of Srinagar city. It has an oval shape and covers an area of approximately 7.23 km^2. Positioned between 34°13′30″ and 34°16′4″N latitudes and 74°30′27″

to 74°32′33″E longitudes, at an altitude of 1585 meters above sea level, this wetland is situated on the flood plains of the Jhelum River, with a maximum depth of 1.2 meters.

Conclusion

The aquatic resources of the Kashmir Himalayas, including rivers, lakes, and wetlands, are invaluable for their ecological, economic, and cultural significance. Major water bodies like Dal Lake, Wular Lake, and the Jhelum River play crucial roles in supporting biodiversity, local livelihoods, and regional hydrology. These resources face significant environmental challenges, such as pollution, habitat degradation, and climate change impacts. Sustainable management and conservation efforts are essential to preserve these ecosystems, ensuring they continue to provide vital benefits for future generations. Enhanced awareness and implementation of effective strategies can contribute to the resilience and sustainability of these precious aquatic systems in the Kashmir Himalayas.

References

Ahmad, N., & Rais, S. (1998). Himalayan glaciers. APH Publishing.

Ahmed, I., Ahmad, Z., & Ahmad, I. (2017). Current status of fish fauna of river Jhelum and dal lake of Kashmir Valley. Bulletin of Pure & Applied Sciences-Zoology, 36(2), 85-92.

Balkhi, M. H. (2007). Fish diversity of Jammu & Kashmir and its conservation. Kashmir speaks. GM Publishers, Kashmir, 104-118.

Bandyopadhyay, J., & Gyawali, D. (1994). Himalayan water resources: ecological and political aspects of management. Mountain Research and Development, 1-24.

Bhat, F. A., & Yousuf, A. R. (2002). Limnological features of a perennial fish pond in the Kashmir Valley. Journal of Himalayan Ecology & Sustainable Development, 1, 19-23.

Bhat, S. A., & Pandit, A. K. (2014). Surface water quality assessment of Wular Lake, a Ramsar site in Kashmir Himalaya, using discriminant analysis and WQI. Journal of Ecosystems, 2014(1), 724728.

Demars, B. O., Russell Manson, J., Ólafsson, J. S., Gíslason, G. M., Gudmundsdóttir, R., Woodward, G. U. Y., & Friberg, N. (2011). Temperature and the metabolic balance of streams. Freshwater Biology, 56(6), 1106-1121.

Directory of Lakes and water-bodies of JandK State (2012). Department of Ecology, Environment and Remote Sensing, Jammu and Kashmir.

Downing, J. (2004). Turbidity monitoring. Environmental instrumentation and analysis handbook, 511-546.

Ganaie, M. I., Dev, A., Mayer, I. A., Ahmed, P., & Ganaie, S. A. (2023). Quality assessment of drinking water sources in the upper Jhelum basin of Kashmir Himalaya, J&K, India. International Journal of River Basin Management, 21(3), 513-524.

Ganie, M. A., & Bhat, F. A. (2008). Status and conservation of Haigam wetland, Kashmir. Journal of Wetlands Ecology, 2, 12-18.

Geelani, S. N. Z., Bhat, B. A., Rasool, S. G., & Dwivedi, S. (2018). Social Appraisal of Wular-Asia's Biggest Fresh Water Lake. Agro-Economist, 5(2), 87-95.

Hajam, S. A., Wani, N. A., & Wani, N. H. (2008). Water quality analysis of Lidder Nalah in Kashmir Valley. Current World Environment, 3(2), 331.

Ishtiaq, H., Fatima, S. N., Khan, J. A., & Khalid, B. (2015). Assessment of Land Cover Change due to Flooding alongside Jhelum and Chenab Rivers. Organizers, 47.

Jamal, S., Ahmad, W. S., Ajmal, U., Aaquib, M., Ashif Ali, M., Babor Ali, M., & Ahmed, S. (2022). An integrated approach for determining the anthropogenic stress responsible for degradation of a Ramsar Site–Wular Lake in Kashmir, India. Marine Geodesy, 45(4), 407-434.

Kullander, S. O., Fang, F., Delling, B. and Ahlander, E. (1999) The fishes of the Kashmir valley. In: River Jhelum, Kashmir valley. Impacts on the aquatic environment. Nyman, L. ed. Swedmar, Goteborgs, Lanstryckeri AB., Swedmar, pp.

Kumar, R., Parvaze, S., Huda, M. B., & Allaie, S. P. (2022). The changing water quality of lakes—a case study of Dal Lake, Kashmir Valley. Environmental monitoring and assessment, 194(3), 228.

Kundangar, M. R. D. (2002). Dal Lake Ecosystem: Conservation Strategies and Problems. ENVIS Bulletin, 10(2), 43-52.

Mehmood, M. A., Shafiq-ur-Rehman, A. R., & Ganie, S. A. (2017). Spatio-temporal changes in water quality of Jhelum River, Kashmir Himalaya. Int J Environ Bioener, 12(1), 1-29.

Mushtaq, F., & Pandey, A. C. (2014). Assessment of land use/land cover dynamics vis-à-vis hydrometeorological variability in Wular Lake environs Kashmir Valley, India using multitemporal satellite data. Arabian Journal of Geosciences, 7, 4707-4715.

Nazir, A., Hussain, S. M., Riyaz, M., & Zargar, M. A. (2024). Microplastic Pollution in Urban-Dal Lake, India: Uncovering Sources and Polymer Analysis for Effective Assessment. Water, Air, & Soil Pollution, 235(2), 89.

Negi, S. S. (1991). Himalayan rivers, lakes, and glaciers. Indus Publishing.

Pandit, A. K. (1999). Freshwater ecosystems of the Himalayas. Proceedings of the Indian National Science Academy, 65(3), 135-146.

Qadri, S. (2022). Fish Fauna of Kashmir Valley and Their Conservational Measures for Sustainable Fish Production. In Biodiversity, Conservation and Sustainability in Asia: Volume 2: Prospects and Challenges in South and Middle Asia (pp. 521-528). Cham: Springer International Publishing.

Qadri, S., Bhat, F. A., Balkhi, M. H., Shah, T. H., Hussain, N., Bhat, B. A. & Aalia, S. (2018). Fish biodiversity, catch composition, distribution pattern of fishes in Wular Lake, and their conservational measures for sustainable fish production. Journal of Pharmacognosy and Phytochemistry, 7(6), 1883-1887.

Raina, A. N. (2002). Geography of Jammu & Kashmir State. Radha Krishan Anand & Co., Pacca Danga, Jammu, 9.

Raina, H. S., & Qadri, M. Y. (1995). Studies on the ecology of Sindh Valley streams in Kashmir. Journal of Research and Development, 4, 48-55.

Rashid, S., & Anees, M. M. (2010). Conservation of Dal Lake: Perspectives and Management. International Journal of Lakes and Rivers, 3(1), 45-52.

Rather, H. A., Ganie, N. A., Gautam, V., & Dar, R. A. (2020) Assessment of the Avian diversity and Habitat preference in Dal Lake Srinagar Kashmir. Wesleyan Journal of Research, 12 (4).

Rather, M. I., Rashid, I., Shahi, N., Murtaza, K. O., Hassan, K., Yousuf, A. R. & Shah, I. Y. (2016). Massive land system changes impact the water quality of the Jhelum River in Kashmir Himalaya. Environmental Monitoring and Assessment, 188, 1-20.

Romshoo, S. A., Altaf, S., Rashid, I., & Dar, R. A. (2018). Climatic, geomorphic, and anthropogenic drivers of the 2014 extreme flooding in the Jhelum basin of Kashmir, India. Geomatics, Natural Hazards and Risk, 9(1), 224-248.

Romshoo, S. A., Rashid, I., Altaf, S., & Dar, G. H. (2020). Jammu and Kashmir State: an overview. Biodiversity of the Himalaya: Jammu and Kashmir State, 129-166.

Shah, M. A., & Pandit, A. K. (2012). Ecological characteristics of Hokersar wetland in Kashmir Himalaya. Proceedings of the National Academy of Sciences, India Section B: Biological Sciences, 82(2), 243-250.

Sharma, V., Mishra, V. D., & Joshi, P. K. (2012). Snow cover variation and streamflow simulation in a snow-fed river basin of the Northwest Himalayas. Journal of Mountain Science, 9, 853-868.

Shukla, A., & Ali, I. (2018). Major river systems of Jammu and Kashmir. The Indian Rivers: Scientific and Socio-economic Aspects, 383-411.

Teng, W., Guoxiang, W., & Qiang, L. (2007). Effects of water turbidity on the photosynthetic characteristics of Myriophyllum spicatum L. Asian Journal of Plant Sciences, 6(5), 773-780.

Wani, M. A., & Pandit, A. K. (2001). Threats to biodiversity in the Kashmir Himalaya. Mountain Research and Development, 21(2), 128-133.

Wei, M., Zhang, N., Zhang, Y., & Zheng, B. (2009). Integrated assessment of river health based on water quality, aquatic life, and physical habitat. Journal of Environmental Sciences, 21(8), 1017-1027.

Wetlands International - South Asia. (2003). Wular Lake: Integrated Management Action Plan. Ministry of Environment and Forests, Government of India.

World Wide Fund for Nature (WWF) India. (2007). Saving Kashmir's Wetlands. WWF-India Report.

Yousuf, A.R., Bhat, F.A. and Mahdi, M.D. (2006). Limnological features of river Jhelum and its important tributaries in Kashmir Himalaya with a note on fish fauna. Journal of Himalayan Ecology and Sustainable Development 1: 37-50.

4

Fish Faunal Diversity of the Trans-Himalayan Region

Zaheer Abass[1], Tasaduq Hussain Shah[1] Karankumar K. Ramteke[2], Rinkesh Nemichand Wanjari[1*] Hudisa Banno[1], Dechen Chuskit[1] and Inab M. Bala[1]

[1]Division of Fisheries Resource Management, Faculty of Fisheries, SKUAST-Kashmir Rangil, Ganderbal, Jammu and Kashmir

[2]Fisheries Resource Harvest and Post-Harvest Management Division, ICAR-Central Institute of Fisheries Education (CIFE), Mumbai, Maharashtra

Abstract

The objective of this chapter is to compile a comprehensive inventory of fish species inhabiting the trans-Himalayan region of Jammu and Kashmir and Ladakh. Through a combination of extensive literature review and field sampling, this study documents the presence of 130 fish species across 10 orders, 23 families, and 58 genera. The majority of these species are classified as Least Concern by the IUCN, though two endemic hill stream fish species, Triplophysa kashmirensis and Glyptothorax kashmirensis, are identified as Critically Endangered. The chapter emphasizes the conservation status of these species and suggests management strategies for their protection and sustainable use. By understanding the fish diversity, distribution patterns, and habitat preferences, effective conservation measures can be devised. The unique biotic zones and abundant cold-water resources of the region underscore its significance as a habitat for freshwater fishes. Despite its importance, research on the fish species in the trans-Himalayan region, particularly in Jammu and Kashmir and Ladakh, remains inadequate. This chapter aims to provide a detailed inventory of the fish species in this region and underscores the need for comprehensive research and conservation efforts to address the alarming decline of riverine fish resources in the trans-Himalayan area.

Keywords: *Fish diversity, IUCN conservation, trans-Himalayan region, cold-water resources, Union Territory*

Introduction

Understanding the diversity, distribution, and habitat preferences of fish species is crucial for devising effective conservation strategies for fisheries resources. Fish distribution patterns are influenced by factors such as dispersal mechanisms, environmental tolerance, and historical factors, while aquatic organisms are directly impacted by variables like climate, land use, geological features, vegetation, topography, and water chemistry fluctuations (Carter *et al.*, 1980; Wiley *et al.*, 1997). Fish represent the most abundant class of vertebrates, comprising nearly half of all vertebrate species globally, with diverse characteristics in terms of size, biology, morphology, and habitat preferences (Kour *et al.*, 2015). Freshwater ecosystems, which account for 68.7% of the world's cold water fisheries, are particularly significant, especially in regions like the Himalayas, where they support vital aquatic life and offer economic opportunities through aquaculture (Fricke *et al.*, 2023). Cold water fish species, predominantly belonging to the Cyprinidontiformes order, are particularly abundant and important in these ecosystems (Fricke *et al.*, 2023). The biological diversity of the planet, including its freshwater ecosystems, has long been a subject of fascination and study (Joshi *et al.*, 2013). India is ranked ninth in the world for freshwater biodiversity, with over 2,500 fish species, 930 of which are unique to freshwater environments. India has been recognized as one of the mega-diversity countries of the world (Mittemeier and Mittemeier, 1997; Jayaram, 2010). The Himalayan region alone contributes approximately 17% of India's fish population (Ghosh, 1997). This chapter aims to comprehensively understand the diversity, distribution, and habitat preferences of fish species in the Kashmir Valley. By examining key factors influencing fish distribution patterns—such as dispersal mechanisms, environmental tolerance, climate, land use, geological features, vegetation, topography, and water chemistry—we seek to develop effective conservation strategies for fisheries resources. Given the significance of freshwater ecosystems, especially in cold-water regions like the Himalayas, this study will shed light on the ecological and economic importance of these habitats. Additionally, by documenting the diversity and distribution of fish species in the Kashmir Valley, this chapter aims to contribute to the broader understanding of freshwater biodiversity, inform conservation policies, and support sustainable management practices for India's rich fisheries resources.

Taxonomic Composition and Species Richness of the Fish Communities

The fish communities of the Kashmir Valley exhibit a rich taxonomic composition and high species richness. Ichthyologists have made significant contributions to our understanding of fish fauna across India, documenting species diversity and distribution through various studies over the years (Hamilton, 1822; Day, 1878; Hora, 1940; Alikunti *et al.*, 1963; Jhingran, 1991; Jayaram, 1999; Sharma *et al.*, 2007; Rahim *et al.*, 2009). While fish inhabit a wide range of aquatic environments, the majority, in terms of species count or abundance, are found in warm environments with moderate temperatures, while others have adapted to cold-water environments. These organisms serve as crucial food resources and indicators of ecological water health. Unfortunately, the deterioration of inland waters is causing a rapid decline in the rich diversity of freshwater fish in the Indian region (Dudgeon *et al.*, 2006; Sarkar *et al.*, 2010, 2012; Lakra *et al.*, 2011).

Identification and Distribution of Major Fish Families and Species

The Kashmir Valley is home to extensive and diverse cold-water/hill resources for fishing (Mahanta and Sarma, 2010). These water bodies exhibit characteristic features such as high-velocity waters, rapid cascades, deep pools, and diverse substrata, supporting rich ichthyofaunal diversity comprising indigenous and exotic fish species (Sehgal, 1999; Mahanta and Sarma, 2010). In India, there are approximately 670 species of freshwater fishes, with 227 of them classified as threatened (NBFGR, 1997). Coldwater fisheries harbor 258 species belonging to 21 families and 76 genera, with significant species diversity observed in various regions, including the North-East Himalaya, west and Central Himalaya, and the Deccan plateau (Vass and Gopakumar, 2002; Vass, 2005; Mahanta, 2010).

Endemism and Unique Evolutionary Adaptations

Besides serving as indicators of water quality and ecosystem health, freshwater fishes are essential in determining the distribution and abundance of other creatures within their ecosystems. However, factors such as impoundment of rivers, discharge of sewage and industrial waste, and habitat degradation have led to freshwater fishes becoming one of the most threatened taxa globally (Darwall and Freyhof, 2016). Many factors endanger the number of fish in freshwater environments in India, including reduced water volume, illegal fishing, unsustainable fishing practices, increased sedimentation, the building of dams and barrages, and water abstraction (Nautiyal *et al.*, 1998; Menon *et al.*, 2000; Langer *et al.*, 2001). Habitat destruction, pollution, species introduction, and commercial exploitation are identified as major contributors to the decline in aquatic biodiversity (Moyle and Leidy, 1992; Kottelat and Whitten, 1996). However, freshwater fish resources in India have received less attention compared to marine species, highlighting the need for comprehensive studies and conservation efforts (Lakra *et al.*, 2010; Pathak *et al.*, 2020). To address the depletion of riverine fish resources, there is a critical need for comprehensive studies on fish fauna inhabiting both lotic and lentic waters, along with effective management strategies to ensure sustainable exploitation and conservation (Sinha *et al.*, 1999; Medhi *et al.*, 2018; Chen *et al.*, 2023).

Ecological Roles and Trophic Relationships

Jammu and Kashmir and Ladakh, possess vast cold-water resources, establishing the region as one of northern India's primary habitats for freshwater fishes (Sarma *et al.*, 2018; Qadri, 2022; Jamwal *et al.*, 2024). These Union Territories (UTs) are endowed with numerous aquatic ecosystems, which are home to economically significant bio resources, including fisheries. The region comprises three distinct biotic zones: Temperate Kashmir, sub-tropical Jammu, and cold arid Ladakh, each characterized by differences in altitude, catchment area, and watershed. The three main rivers in the Indus basin—the Jhelum, the Chenab, and the plentiful snow- and ice-melt waters—are vital to the approximately 300 million people who live upstream and downstream in the basin by providing food, energy, and water. The area has over 3,650 wetlands and other bodies of water, and it is well-known for its freshwater lakes and wetlands (Romshoo *et al.*, 2020). The present status and distributional patterns of fish fauna in the Himalayan region remain poorly unexplored, highlighting the need for comprehensive research on fish species in the Union Territory (UT) of Jammu and Kashmir, and Ladakh. Thus, the present study aims to provide a detailed inventory and checklist of fish species in the trans-Himalayan region.

Fish diversity in the Trans-Himalayan region

Kashmir Valley

Renowned for its abundant fishery resources, the Valley of Kashmir, nestled in the northern Himalayas of India, presents significant opportunities for fish production both for local consumption and export, accounting for 31% of the country's total cold-water fish production (Qayoom *et al.*, 2019). The region's extensive water resources, including lakes, wetlands, ponds, streams, springs, and rivers, serve as prime habitats for cold-water fisheries, contributing significantly to the development of the fisheries sector (Qadri, 2022). The rivers of the Kashmir Valley host a diverse range of indigenous fish species such as *Schizothorax* spp., *Glyptothorax* spp., and *Triplophysa* spp. Additionally, the region is globally acclaimed for its exotic brown trout (*Salmo trutta fario*) and rainbow trout (*Onchorhynchus mykiss*). Introduced to the Valley in the early 20th century, these trout species have become integral to the region's sport fishery, attracting visitors from around the world. Schizothorax, commonly referred to as snow trout or Himalayan trout, holds significant importance as a key food fish in the Kashmir Himalaya, thriving in the network of snow-fed and spring-fed cool water rivers and streams throughout the region. Since Heckel's pioneering report on Kashmir's fish in 1838, several researchers have contributed to documenting the fish fauna of the valley (Day, 1876; Silas, 1960; Das and Subla, 1964; Das, 1965; Yousuf, 1996; Enderlin and Yousuf, 1999; Kullander *et al.*, 1999; Balkhi, 2007; Sarkar *et al.*, 2010; Lakra *et al.* 2010; Vijaylakshmi *et al.*, 2010). Reported numbers of fish species by these researchers vary between 18 and 42, possibly due to synonymy or uncertainty in identification. In this study based on field works and literature survey we have observed that the mighty mahseer (*Tor putitora*) is completely absence in the stretch of Jhelum River, in Kashmir Valley, due to the construction of Mangla Dam and Uri Hydroelectric Power Project in situated down-stream in the Pakistan and Kashmir (India) respectively. The absence of a fish ladder in this dam effectively restricts the anadromous fish's migration path. The fish had previously found its spawning grounds in the Kashmir valley, India (Nyman, 1995). The lower Jhelum Hydroelectric Project, constructed on the River Jhelum, has had significant detrimental environmental effects, particularly impacting the aquatic biodiversity of the river. This stretch of the river was once home to 18 fish species belonging to 12 genera, but since the project's initiation, various adverse effects have been observed, notably impacting the population of the formidable mighty mahseer, (*T. putitora*), now classified as an endangered species. Present survey revealed the occurrence of 22 fish species belonging to 4 orders, 6 families and 16 genera. Four orders namely, Cypriniformes, Siluriformes, Salmoniformes, and Cyprinodontiformes. In Cypriniformes, the survey identified species from three families: Cyprinidae (13 species), Nemacheilidae (3 species), and Botiidae (1 species). Siluriformes had two species from the Sisoridae family, while Salmoniformes had two species from Salmonidae. Moreover, Cyprinodontiformes contained one species from the Poeciliidae family in the Kashmir Valley (Table 1 and Fig. 1).

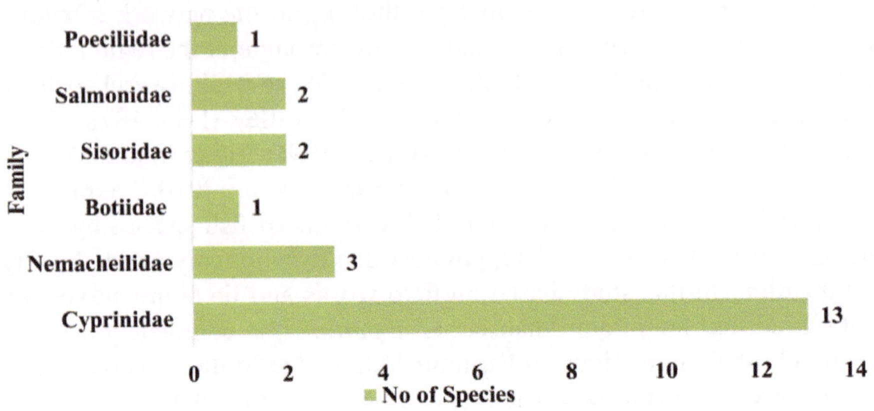

Fig. 1 Number of fish species in each family (Kashmir Valley).

Table 2. List of ichthyofaunal diversity in Kashmir Valley.

Sr. No	Order	Family	Scientific Name	Common/ Local Name	IUCN Status
1	Cyprini-formes	Cyprinidae	*Schizothorax niger* (Heckel,1838)	Alegad/snow trout	Least Concern
2			*Schizothorax labiatus* (McClelland, 1842)	Kunar/Chush	Not Evaluated
3			*Schizothorax esocinus* (Heckel, 1838)	Chirruh	Vulnerable
4			*Schizothorax plagios-tomus* (Heckel, 1838)	Snow carp/ Khont	Vulnerable
5			*Schizothorax curvifrons* (Heckel, 1838)	Sattar gad / sattar snow trout	Least Concern
6			*Cyprinus carpio* var. *communis* (Linnaeus, 1758)	Punjab gad	Vulnerable
			Cyprinus carpio var. *specularis* (Lacepede, 1758)	Punjab gad	Vulnerable
7			*Carassius carassius* (Linnaeus, 1758)	Gang gad/ Crucian carp	Least Concern
8			*Pethia conchonius* (Hamilton, 1822)	Safaid bach/ Rosy barb	Least Concern
9			*Bangana diplostoma* (Heckel, 1838)	Roppat	Least Concern
10			*Crossocheilus diplochilus* (Heckel, 1838)	Tethur	Not Evaluated

11			*Diptychus maculatus* (Steindachner, 1866)	Scaly osman	Not Evaluated
12			*Hypophthalmichthys molitrix* (Valenciennes, 1844)	Silver carp	Near Threatened
13			*Ctenopharyngodon idella* (Valenciennes, 1844)	Grass carp	Least Concern
14		Botiidae	*Botia birdi* (Chaudhuri, 1909)	Birdi loach	Not Evaluated
15		Nemacheili-dae	*Paraschistura punjabensis* (Hora, 1923)		Vulnerable
16			*Triplophysa marmorata* (Heckel, 1838)	Ara gurun	Least Concern
17			*Triplophysa kashmirensis* (Hora, 1922)	Ara gurun	Critically En-dangered
18	Siluriformes	Sisoridae	*Glyptosternon reticulatum* (McClelland, 1842)	Nayid	Least Concern
19			*Glyptothorax kashmirensis* (Hora, 1923)	Nayid	Critically En-dangered
20	Salmoni-formes	Salmonidae	*Salmo trutta fario* (Linnaeus, 1758)	Brown Trout	Least Concern
21			*Oncorhynchus mykiss* (Walbaum, 1792)	Rainbow Trout/Trooth	Not Evaluated
22	Cyprinodonti-formes	Poeciliidae	*Gambusia holbrooki* Girard, 1859	Eastern mosquito-fish	Least Concern

Jammu Region

The ecological and geographical diversity of Jammu province underwent extensive exploration, initially documented by Das and Nath (1965, 1966), who recorded 20 fish species from the Chenab drainage system. Subsequent researchers further investigated the region's fish diversity across various areas (Das and Nath, 1971; Malhotra *et al.*, 1975; Malhotra and Dutta, 1976; Dutta and Malhotra, 1984; Jyoti *et al.*, 2006; Balkhi, 2007). Rathore and Datta (2015) expanded this knowledge by documenting 42 fish species in the River Ujh, a tributary of the River Ravi. Additionally, Datta *et al.* (2015) conducted a comprehensive survey of ichthyofauna in Udhampur district, identifying 76 fish species spanning 6 orders, 14 families, and 45 genera. Notably, they identified *Glyptothorax punjabensis* (Mirza and Kashmiri) as a new record for the country and recognized *Barilius radiolatus* (Gunther), *Botia lohachata* (Chaudhari), and *Botia dario* (Ham. Buch.) as new records for the UT. Their survey also included stocked species introduced by the state fishery department. Further contributing to the repository, Datta *et al.* (2016) conducted a survey of fish diversity within Wajoo Nullah in the Kathua district, uncovering 64 fish species distributed across 7 orders, 17 families, and 42 genera. Notably, they identified *Gudusia chapra* and *Parambassis bacails* as new additions to the Jammu and Kashmir state fish records. Building upon these previous studies, Bhat *et al.* (2020) reported a total of 105 fish species in the Jammu region as part of a state-wide fishery survey initiated by the Indian Council of Agricultural Research (ICAR). Their study primarily focused on assessing fish diversity in the Jammu region. In the present survey Jammu region has revealed the occurrence of 113 fish species, belonging to 8 orders, 22 families, and 58 genera. These species include representatives from Cypriniformes, Siluriformes, Osteoglossiformes, Clupeiformes, Perciformes, Beloniformes, Synbranchiformes, and Salmoniformes. Out of these, 66 species belonging to the order Cypriniformes (comprising 56 from the family Cyprinidae, 6 from Nemacheilidae, 3 from Botiidae and 1 from Cobitidae), two species to Salmoniformes (family Salmonidae), and 25 species to Siluriformes (12 from Sisoridae ,5 from Bagridae, 3 form Siluridae, 3 from Schilibeidae, 1 from Heteropneustidae and 1 from Amblycptidae,), 2 species belong from Osteoglossiformes (family Notopteridae), 12 species from order perciformes (6 from family Channidae, 2 from family Ambassidae, 1 in the each from family Osphronemidae, Nandidae, Badidae, Gobiidae). 1 species from the order Clupeformes (Family Clupeidae), 1 species from the order Beloniformes (Belonidae), 4 species from the order Synbranchiformes (3 from the family Mastacembelidae and 1 from the family Synbranchidae). Notably, Cypriniformes displayed the highest diversity with 56 species, followed by Osteoglossiformes with 12 species. This comprehensive survey underscores the remarkable richness of fish biodiversity in the Jammu region (Table 2 and Fig. 2).

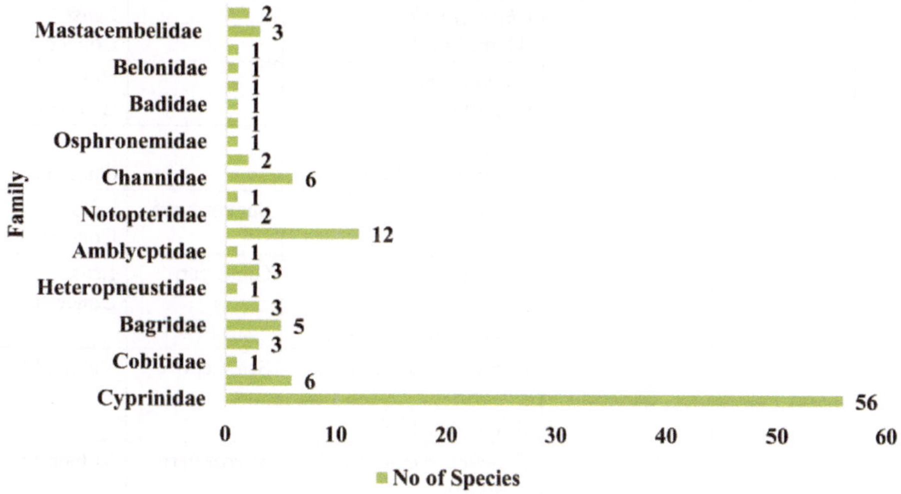

Fig. 2 Number of fish species in each family (Jammu region)

Table 3. List of ichthyofaunal diversity of Jammu region

Sr. No	Order	Family	Scientific Name	Common/ Local Name	IUCN Status
1	Cyprini-formes	Cyprinidae	*Aspidoparia morar* (Hamilton, 1822)		Least Concern
2			*Amblypharyngodon gudigurhi* (Malhotra & Singh Dutta, 1975)		Least Concern
3			*Amblypharyngodon mola* (Hamilton, 1822)	Mola carplet	Least Concern
4			*Barilius bendelisis bendelisis* (Hamilton, 1807)		Least Concern
5			*Barilius vagra* (Hamilton, 1822)		Least Concern
6			*Barilius barila* (Hamilton, 1822)		Least Concern
7			*Bangana diplostoma* (Heckel, 1838)		Least Concern
8			*Cirrhinus mrigala* (Hamilton, 1822)		Least Concern
9			*Cirrhinus reba* (Hamilton, 1822)	Reba carp	Least Concern

10			*Chela cachius* (Hamilton, 1822)		Least Concern
11			*Chela atpar* (Hamilton, 1822)		Not Evaluated
12			*Laubuka laubuca* (Hamilton, 1822)		En-dangered
13			*Garra gotyla* (Gray, 1830)	Sucker head	Least Concern
14			*Ctenopharyngodon idella* (Valenciennes, 1844)	Grass carp	Least Concern
15			*Cyprinus carpio* var. *communis* (Linnaeus, 1758)	Scale carp	Vulnerable
			Cyprinus carpio var. *specularis* (Lacepede, 1758)	Mirror carp	Vulnerable
16			*Devario devario* (Hamilton, 1822)		Least Concern
17			*Danio rerio* (Hamilton, 1822)	Zebra danio	Least Concern
18			*Esomus danricus* (Hamilton, 1822)	Flying barb	Least Concern
19			*Garra lamta* (Hamilton, 1822)		Least Concern
20			*Garra variabilis* (Heckel, 1843)	Variable garra	Least Concern
21			*Hypophthalmichthys molitrix* (Valenciennes, 1844)	Silver carp	Near Threatened
22			*Crossochilus latius diplochilus* Hamilton		Least Concern
23			*Labeo catla* (Hamilton, 1822)	Catla	Least Concern
24			*Labeo rohita* (Hamilton, 1822)	Roho labeo	Least Concern
25			*Labeo dyocheilus* (McClelland, 1839)		Least Concern
26			*Labeo boggut* (Sykes, 1839)		Least Concern
27			*Labeo bata* (Hamilton, 1822)	Bata	Least Concern

28			*Labeo calbasu* (Hamilton, 1822)	Orangefin labeo	Least Concern
29			*Labeo gonius* (Hamilton, 1822)	Kuria Labeo	Not Evaluated
30			*Labeo dero* (Hamilton, 1822)		Least Concern
31			*Labeo pangusia* (Hamilton, 1822)		Near Threatened
32			*Osteobrama cotio* (Hamilton, 1822)		Least Concern
33			*Oxygaster bacaila* (Hamilton, 1822)		Least Concern
34			*Pethia conchonius* (Hamilton, 1822)	Rosy barb	Least Concern
35			*Pethia ticto* (Hamilton, 1822)	Ticto barb	Least Concern
36			*Systomus sarana* (Hamilton, 1822)	Olive barb	Least Concern
37			*Cabdio morar* (Hamilton, 1822)		Least Concern
38			*Rasbora rasbora* (Hamilton, 1822)		Least Concern
39			*Salmostoma bacaila* (Hamilton, 1822)	Large razorbelly minnow	Least Concern
40			*Salmophasia punjabensis* (Day, 1872)		Not Evaluated
41			*Schizothorax plagiostomus* (Heckel, 1838)	Snow carp/ Khont	Vulnerable
42			*Schizothorax labiatus* (McClelland, 1842)	Kunar/Chush	Not Evaluated
43			*Schizothorax esocinus* (Heckel, 1838)	Chirruh	Vulnerable
44			*Schizothorax curvifrons* (Heckel, 1838)	Sattar snowtrout	Least Concern
45			*Schizothorax progastus* (McClelland, 1839)	Dinnawah snowtrout	Least Concern
46			*Schizothorax richardsonii* (Gray, 1832)	Snow trout	Vulnerable

47			*Tor putitora* (Hamilton, 1822)	Putitor mahseer/ Golden mahseer	Endangered
48			*Tor tor* (Hamilton, 1822)		Endangered
49			*Tor khudree* (Sykes, 1839)	Deccan mahseer	Least Concern
50			*Securicula gora* (Hamilton, 1822)		Least Concern
51			*Barilius modestus* Day, 1872		Not Evaluated
52			*Puntius sophore* (Hamilton, 1822)		Least Concern
53			*Puntius chola* (Hamilton, 1822)		Least Concern
54			*Puntius chola* (Hamilton, 1822)	Swamp barb	Least Concern
55			*Puntius terio* (Hamilton, 1822)	Onespot barb	Least Concern
56			*Barilius radiolatus* (Günther, 1868)		Data Deficient
57		Nemacheilidae	*Acanthocobitis botia* (Hamilton, 1822)	Mottled loach	Least Concern
58			*Triplophysa kashmirensis* (Hora, 1922)		Critically Endangered
59			*Triplophysa yasinensis* (Alcock, 1898)		Not Evaluated
60			*Paraschistura punjabensis* (Hora, 1923)		Vulnerable
61			*Nemacheilus corica* (Hamilton, 1822)		Least Concern
62			*Paraschistura parshadi* (Hora, 1933)		Vulnerable
63		Botiidae	*Botia almorhae* Gray, 1831	Almorha loach	Least Concern
64			*Botia lohachata* Chaudhuri, 1912	Reticulate loach	Least Concern
65			*Botia dayi* (Hora, 1932)		Not Evaluated

66		Cobitidae	*Lepidocephalichthys guntea* (Hamilton, 1822)	Guntea loach	Least Concern
67	Siluriformes	Bagridae	*Mystus bleekeri* (Day, 1877)	Day's mystus	Least Concern
68			*Mystus cavasius* (Hamilton, 1822)	Gangetic mystus	Least Concern
69			*Mystus vittatus* (Bloch, 1794)	Striped dwarf catfish	Least Concern
70			*Rita rita* (Hamilton, 1822)	Ritha	Least Concern
71			*Sperata seenghala* (Sykes, 1839)		Least Concern
72		Siluridae	*Ompok bimaculatus* (Bloch, 1794)		Near Threatened
73			*Wallago attu* (Bloch & Schneider, 1801)		Near Threatened
74			*Ompok pabda* (Hamilton, 1822)	Pabdah catfish	Near Threatened
75		Heteropneusti-dae	*Heteropneustes fossilis* (Bloch, 1794)	Singhi	Least Concern
76		Schilibeidae	*Clupisoma garua* (Hamilton, 1822)	Garua bachcha	Least Concern
77			*Eutropiichthys vacha* (Hamilton, 1822)	Batchwa vacha	Least Concern
78			*Pachypterus atherinoides* (Bloch, 1794)	Indian potasi	Least Concern
79		Amblycptidae	*Amblyceps mangois* (Hamilton, 1822)	Indian torrent catfish	Least Concern
80		Sisoridae	*Bagarius bagarius* (Hamilton, 1822)	Goonch	Vulnerable
81			*Gagata cenia* (Hamilton, 1822)	Indian gagata	Least Concern
82			*Glyptothorax pectinopterus* (McClelland, 1842)	River cat	Least Concern
83			*Glyptothorax kashmirensis* (Hora, 1923)	Nayid	Critically Endangered
84			*Glyptothorax stolickae* (Steindachner, 1867)		Least Concern

85			*Glyptosternon reticulatum* (McClelland, 1842)	Nayid	Least Concern
86			*Glyptothorax botius* (Hamilton, 1822)		Least Concern
87			*Glyptothorax punjabensis* Mirza & Kashmiri, 1971		Endangered
88			*Glyptothorax cavia* (Hamilton, 1822)		Least Concern
89			*Glyptothorax garhwali* Tilak, 1969		Least Concern
90			*Glyptothorax telchitta* (Hamilton, 1822)		Least Concern
91			*Glyptothorax indicus* Talwar, 1991		Least Concern
92	Osteoglossi-formes	Notopteridae	*Notopterus notopterus* (Pallas, 1769)	Bronze feather-back	Least Concern
93			*Chitala chitala* (Hamilton, 1822)	Clown knifefish	Near Threatened
94	Clupe-formes	Clupeidae	*Gudusia chapra* (Hamilton, 1822)	Indian river shad	Least Concern
95	Perciformes	Channidae	*Channa punctata* (Bloch, 1793)	Spotted snakehead	Least Concern
96			*Chanda nama* Hamilton, 1822	Elongate glass-perchlet	Least Concern
97			*Channa marulius* (Hamilton, 1822)	Great snakehead	Least Concern
98			*Channa gachua* (Hamilton, 1822)	Red seam snakehead	Least Concern
99			*Channa orientalis* Bloch & Schneider, 1801	Walking snakehead	Vulnerable
100			*Channa striata* (Bloch, 1793)	snakehead	Least Concern
101		Ambassidae	*Parambassis ranga* (Hamilton, 1822)	Indian glassy fish	Least Concern
102			*Parambassis baculis* (Hamilton, 1822)	Himalayan glassy perchlet	Least Concern

103		Osphronemidae	*Trichogaster fasciata* Bloch & Schneider, 1801	Banded gourami	Least Concern
104		Nandidae	*Nandus nandus* (Hamilton, 1822)	Gangetic leaffish	Least Concern
105		Badidae	*Badis badis* (Hamilton, 1822)		Least Concern
106		Gobiidae	*Glossogobius giuris* (Hamilton, 1822)	Tank goby	Least Concern
107	Beloniformes	Belonidae	*Xenentodon cancila* (Hamilton, 1822)		Least Concern
108	Synbranchiformes	Synbranchidae	*Monopterus cuchia* (Hamilton, 1822)	Cuchia	Least Concern
109		Mastacembelidae	*Mastacembelus armatus* (Lacepède, 1800)	Zig-zag eel	Least Concern
110			*Macrognathus aral* (Bloch & Schneider, 1801)	One-stripe spinyeel	Least Concern
111			*Macrognathus pancalus* Hamilton, 1822	Barred spiny eel	Least Concern
112	Salmoniformes	Salmonidae	*Salmo trutta fario* (Linnaeus, 1758)	Brown Trout	Least Concern
113			*Oncorhynchus mykiss* (Walbaum, 1792)	Rainbow Trout/Trooth	Not Evaluated

Ladakh Region

In Ladakh, the aquatic realm has been extensively studied, resulting in the documentation of various fish species. Tilak (1990) provided a detailed account of 22 fish species, including representatives from Cyprinidae, Cobitidae, and Sisoridae. Notably, two new species within the *Noemacheilus* genus, *N. drassensis* and *N. shehensis*, were discovered. Sivakumar (2008) expanded the list to 32 species, focusing on the Indus Shyok and Zanskar rivers and incorporating five exotic fish species found within the Indus catchment. However, a subsequent analysis revealed that the list by Sivakumar (2008) may have been inflated due to the inclusion of synonyms and exotic species bred in enclosed fish farms. Bhat *et al.* (2020) conducted a study documenting 15 fish species in Ladakh, classifying them into Cypriniformes (Cyprinidae and Balitoridae families), Siluriformes (Sisoridae family), and Salmoniformes (Salmonidae family). Another study by Kumar *et al.* (2020) recorded 26 fish species, including indigenous and exotic varieties from four families. The Indus River exhibited the highest diversity, with 23 species, followed by the Shyok River with 12 species and the Zanskar River with the lowest at ten species. It is worth noting that commercial fishing is not practiced in Ladakh's waters. While religious considerations discourage fish consumption among the majority of the population in Leh and Kargil districts, there is a noticeable demand for fish from the local populace and security forces. This has led to the establishment of several fish farms in the region to meet the demand for fish products. The reported fish diversity in the region of Ladakh has led to some ambiguity. Various authors have provided varying numbers of fish species found in this area. In this study actual count reported that there are 25 fish species present in the cold arid Ladakh region. However, Kumar *et al.* (2020) documented a total of 26 fish species, including two strains of the common carp, namely *specularis* and *communis*. It is important to note that both strains of the common carp belongs to the same species but different strains of the same species. Therefore, the total number of species documented in that study appears to be higher compared to the present study. Present survey revealed the occurrence of 25 fish species belonging to 3 orders, 5 families and 13 genera in the Ladakh region. These species belong to three orders: Cypriniformes, Siluriformes, and Salmoniformes. Among these orders, Cypriniformes is the most diverse, with representatives from three families: Cyprinidae (12 species), Botiidae (1 species), and Nemacheilidae (9 species). Siluriformes is represented by one family, Sisoridae (1 species), while Salmoniformes is represented by Salmonidae (2 species) (Table 3 and Fig. 3). Figure 4 depicting overall fish diversity IUCN status in the Trans-Himalayan region.

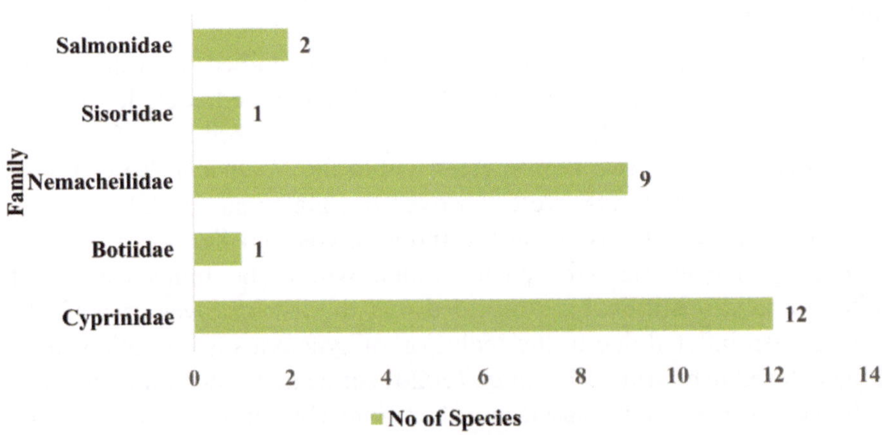

Fig. 3 Number of fish species in each family (Ladakh region).

Table 3. List of ichthyofaunal diversity of Ladakh region.

Sr. No	Order	Family	Scientific Name	Common/ Local Name	IUCN Status
1	Cyprini-formes	Cyprini-dae	*Schizothorax richardsonii* (Gray, 1832)	Snow trout	Vulnerable
2			*Schizothorax progastus* (McClelland, 1839)	Dinnawah snowtrout	Least Concern
3			*Schizothorax labiatus* (McClelland, 1842)		Not evaluated
4			*Schizothorax esocinus* (Heckel, 1838)		Vulnerable
5			*Schizothorax plagiostomus* (Heckel, 1838)	Nya	Vulnerable
6			*Schizothorax curvifrons* (Heckel, 1838)		Least Concern
7			*Schizopygopsis stoliczkae* (Steindachner, 1886)	False osman	Not Evaluated
8			*Diptychus maculatus* (Steindachner, 1866)	Scaly osman	Least Concern
9			*Ptychobarbus conirostris* (Steindachner, 1866)	Indus snowtrout	Least Concern
10			*Ptychobarbus kaznakovi* Nikolskii, 1903		Not Evaluated
11			*Crossocheilus diplocheilus* (Heckel, 1838)		Not Evaluated
12			*Cyprinus carpio* Var. *communis* (Linnaeus, 1758)	Scale carp	Vulnerable
			*Cyprinus carpio*Var. *specularis* (Lacepede, 1758)	Mirror carp	Vulnerable
13		Nemacheili-dae	*Acanthocobitis botia* (Hamilton, 1822)	Mottled loach	Least Concern
14			*Paraschis turamontana* (McClelland, 1838)		Not Evaluated
15			*Paraschis turaprashari* (Hora, 1933)		Not Evaluated
16			*Schistura fascimaculata* (Mirza & Nalbant, 1981)		Vulnerable
17			*Schistura arifi* (Mirza Banarescu, 1981)		Least Concern
18			*Triplophys ayasinensis* (Alock, 1898)		Not Evaluated

19			*Triplophys gracilis* (Day, 1877)		Not Evaluated
20			*Triplophys tenuicauda* (Steindachner, 1866)		Data Deficient
21			*Triplophys amicrops* (Steindachner, 1866)		Least Concern
22		Botiidae	*Botia dayi* (Hora, 1932)	Hora loach	Not Evaluated
23	Siluri-formes	Sisoridae	*Glyptosternon reticulatum* (McClelland, 1842)	Nayid	Least Concern
24	Salmoni-formes	Salmonidae	*Oncorhynchus mykiss* (Walbaum, 1792)	Rainbow Trout	Not Evaluated
25			*Salmo trutta fario* (Linnaeus, 1758) (Ex)	Brown Trout	Least Concern

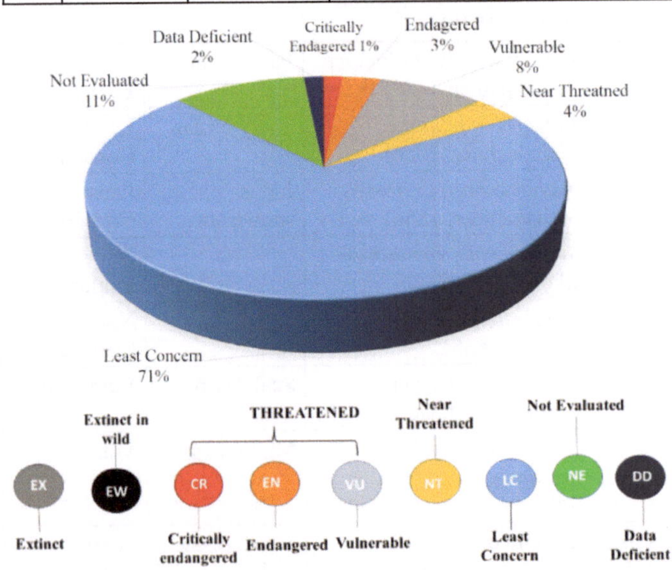

Fig. 4 IUCN status of overall fish diversity

Strategies for Sustainable Management and Conservation of Fish Biodiversity

The fishery resources of Jammu and Kashmir face several challenges, including environmental degradation, pollution, and habitat destruction. The construction of dams, reservoirs, and hydropower projects has significantly altered the natural flow regimes and habitats of the rivers, adversely affecting fish populations. Climate change and global warming also pose a significant threat to the region's aquatic ecosystems, leading to changes in water temperature, flow patterns, and habitat availability (Bhat *et al.*, 2020). Furthermore, the introduction of exotic fish species has had mixed effects on the native fish fauna, sometimes leading to competition, predation, and hybridization with indigenous species. Effective management and conservation strategies are essential to address these issues and ensure the sustainability of the fisheries resources in the Kashmir Valley.

Conclusion

The diverse fish fauna of the Kashmir Valley reflects the region's unique ecological conditions and evolutionary history. While the valley hosts a variety of indigenous and exotic species, it faces significant challenges from environmental degradation, habitat loss, and climate change. Comprehensive research, effective management, and conservation strategies are essential to protect and sustain the fish biodiversity of this ecologically and economically significant region. By understanding the intricate dynamics of fish communities and their habitats, we can develop informed policies to ensure the long-term sustainability of fisheries resources in the Kashmir Valley.

References

Alikunti KH, Sukumaran KK, Parameshwaran S. 1963. Induced breeding of the Chinese carps *Ctenopharyngodon idellus* (C and V) and *Hypopthalmicthyes molitrix* (C and V) in ponds at Cuttack, India. Current Science 32:103–126

Awas M, Ahmed I, Ahmad SM. 2023. Habitat ecology and current status of the fish fauna of River Poonch of Pir Panjal Himalayan region of Jammu and Kashmir, India. Tropical Ecology 64(4): 655-670.

Azam S, Saha AKM, Asadujjaman KR, Minar MH. 2014. Fishing gears and crafts commonly used at Hatiya Island: A coastal region of Bangladesh. Asian Journal of Agricultural Research 8 (1):51-58.

Balkhi MH. 2007. Fish diversity of Jammu & Kashmir and its conservation. Kashmir speaks. GM Publishers, Kashmir 104-118.

Bhat FA, Yousuf AR, Balkhi MH. 2020. Diversity of fishes in Jammu and Kashmir state. Biodiversity of the Himalaya: Jammu and Kashmir State 859-887.

Carter JCH, Dadswell MJ, Rof JC, Sprules WG. 1980. Distribution of zoogeography of planktonic crustaceans and dipterans in glaciated eastern North America. Canadian Journal of Zoology 58:1355–1387.

Chen Q, Li Q, Lin Y, Zhang J, Xia J, Ni J, Cooke SJ, Best J, He S, Feng T, Chen Y. 2023. River damming impacts on fish habitat and associated conservation measures. Reviews of Geophysics 61(4)

Dar SA, Desai AY, Rather AM, Sayani AN, Parmar ER, Arjamand S, Chesti A. 2014. Fishing gears operated along the Wular Lake, Jammu & Kashmir, India. Ecology Environment and Conservation 2:45-47.

Darwall WR, Freyhof JÖRG. 2016. Lost fishes, who is counting. The extent of the threat to freshwater fish biodiversity. Conservation of freshwater fishes 1-36.

Das SM, Nath S. 1965. The ichthyofauna of Poonch valley (J&K). Kashmir Sci 2(1-2): 149-155.

Das SM, Nath S. 1966. The ichthyofauna of Jammu with their ecology. Proc. 53rd Ind. Sci. Congr 374-375.

Das SM, Nath S. 1971. A revision of the fishes from Jammu provinces Kashmir.

Das SM. 1965. A revision of the fish species inhabiting Kashmir Province. Kashmir Sci 2: 13, 19.

Day F. 1878. The fishes of India, reproduced in 1958. Mlliam Dowson and Sons Ltd., London, p 1.

Dudgeon D, Arthington AH, Gessner MO, Kawabata ZI, Knowler DJ, Lévêque C, Sullivan, CA. 2006. Freshwater biodiversity: importance, threats, status and conservation challenges. Biological reviews 81(2): 163-182.

Dutta SPS, Malhotra YR. 1984. An upto date checklist and a key to the identification of fishes of Jammu. Jammu Univ. Review 2: 65-92.

Dutta SPS. 2015. Survey and systematic analysis of Fish fauna of erstwhile Udhampur District, Jammu region, with new records of Barilius radiolatus Gunther, Botia lohachata Chaudhari, Botia dario (Ham. Buch.) and Glyptothorax punjabensis Mirza and Kashmiri. Environment Conservation Journal 16(3): 39-47.

Dutta SPS. 2016. Fish fauna of Wajoo nullah, an important tributary of the river Ravi in Kathua District, Jammu Region, Jammu and Kashmir State, India. Journal of Applied and Natural Science 8(2): 1087-1089.

Fricke R, Eschmeyer WN, Van der Laan R.2023. Eschmeyer's catalog of fishes: genera, species, references. http://researcharchive.calacademy.org/research/ichthyology/ catalog/ fishcatmain.asp. Accessed on: 2023-01-20

Ghosh AK. 1997. Himalayan fauna with special reference to endangered and endemic species. In: Dhar U (ed) Himalayan Biodiversity: Action plan, GB Pant Institute of Himalayan Environment & Development, Kosi-Katarmal, Almora, pp 53–59

Hameed MS, Boophendranath MR. 2000. Modern Fishing Gear Technology, Das Publishing House, Delhi. 186p.

Hamilton BF. 1822. An account of fishes found in the river Ganges and its branches. Edinburg London, p 405

Hora SL. 1940. On a collection of fish from head waters of the Mahanadi River, Raipur district, Central province. Rec India Mus 42:365–372

Jamwal A, Phulia V, Mushtaq ST. 2024. Predictable Threats to Coldwater Fisheries from the Unpredictable Nature of Climate Change: An Indian Perspective. In Coldwater Fisheries and Aquaculture Management (pp. 407-418). Apple Academic Press.

Jayaram KC. 2010. The Freshwater Fishes of the Indian Region. 2nd Edition, Narendra Publishing House, Delhi, 616 p

Jayaram KC. 1999. The Freshwater Fishes of the Indian Region. Narendra Publishing House, New Delhi, p 551

Jhingran VG. 1991. Fish and Fisheries of India. Revised and enlarged 3rd edition. Hindustan Publishing Corporation, New Delhi, India, pp 125–128

Jyoti MK, Sharma K, Gupta, K, and Baba DI. 2006. Aquatic biodiversity: A review of fresh water flora and fauna of J&K state. In Proceeding National Symposium: status of coldwater fishes with reference to fragile Himalayan aquatic ecosystems. Classic Printers, Bari Brahmna (pp. 258-292).

Khanna SS, Singh HR. 2003. A text book of Fish biology and Fisheries. Narendra Publishing House, Delhi, India 524.

Kottelat M, Whitten T. 1996. Freshwater Biodiversity in Asia with Special Reference to Fish. World Bank Technical Paper No. 343, Washington, 59 pp

Kour H, Sharma KK, Sharma A. 2015. Icthyofaunistic survey of some lotic water bodies of R. S. Pura, tehsil, Jammu. Journal of International Academic Research for Multidisciplinary, 3: 2320-5083.

Kullander SO, Fang F, Delling B, Ahlander E. 1999. The fishes of the Kashmir Valley. River Jhelum, Kashmir Valley. Impacts on the aquatic environment. Swedmar, Göteborg, 99-167.

Kumar P, Ingole NA, Ganie PA, Posti R, Saxena AK, Ahmad Z. 2020. Assessment of fishery resources in Trans Himalayas of union territory Ladakh region (India) using geographical information system. Journal of Experimental Zoology India 23(2).

Lakra WS, Sarkar UK, Dubey VK, Sani R, Pandey A. 2011. River inter linking in India: status, issues, prospects and implications on aquatic ecosystems and freshwater fish diversity. Reviews in Fish Biology and Fisheries 21: 463-479.

Lakra WS, Sarkar UK, Gopalakrishnan A, Kathirvelpandian A. 2010. Threatened freshwater fishes of India. National Bureau of Fish Genetic Resources, Indian Council of Agricultural Research India. 25p.

Langer RK, Ogale SN, Ayyappan S. 2001. Mahseer in Indian subcontinent-a bibliography. Central Institute of Fisheries Education, Mumbai

Mahanta PC, Sarma D. 2010. Coldwater Fisheries Management. DCFR, ICAR, Bhimtal – 263 136, Distt. Nainital (Uttarakhand) India. P. 1-451.

Malhotra YR, Dutta SPS. 1976. On two new fish species from Jammu along with a checklist of fishes inhabiting Jammu division of J & K State. India. Proc. Nat Acad Sci India 40: 158-161.

Malhotra YR, Jyoti MK, Dutta SPS. 1975. An aid to the identification of fishes found in Jammu Division of J&K State. J. Univ. Review 5(8): 50-66.

Malik R, Abubakr A, Hussain N. 2018. Indigenous fish attractants, fishing methods, gears and storage: a study in the fishing community of Wular Lake of district Bandipora of Kashmir valley. Journal of Pharmacognosy and Phytochemistry 7(4):1751-1755.

Medhi K, Paul T, Borah AK, Bhattacharjya BK, Debnath D, Yengkokpam S, Kuberan G. 2018. Fish Stock Depletion in Indian Inland Open Waters: Present status and Approaches for Sustainable use, Conservation and Management. Aquaculture Times, 4(5), pp.26-32.

Menon AGK, Singh HR, and Kumar N. 2000. Present eco status of cold-water fish and fisheries. In: Singh HR, Lakra WS (eds) Coldwater Fish and Fisheries, Narendra Publishing House, New Delhi. pp 1–36.

Mittermeier RA, Mittemeier CG. 1997. Megadiversity: Earth's Biologically Wealthiest Nation. In McAllister, D.E., Hamilton, A.L. and Harvery, B. (eds.), Global Freshwater Biodiversity. Sea Wind, Cemex, Mexico City, 11: 1-140.

Moyle PB, RA Leidy 1992. Loss of biodiversity in aquatic ecosystems: evidence from fish faunas. In: Fiedler, P.L. and S.K. Jain (eds.). Conservation Biology: The Theory and Practice of Nature Conservation, Preservation and Management pp. 127-169

Mundy P, Campton JL. 1991. Indigenous communication and indigenous knowledge. Dev. Commun. Rep 74(3): 1-4.

Nautiyal P, Bhatt JP, Rawat VS, Kishor B, Nautiyal R, Singh HR. 1998. Himalayan Mahseer: Magnitude of commercial fishery in Garhwal hills. Fish Gen. Biodiversity Conserv., Netcon Publ 5: 107–14.

NBFGR. 1997. Conservation assessment and management plan for freshwater fishes of India. Technical report. National Bureau of Fish Genetic Resources, Lucknow. 46 pp.

Pathak AK, Sarkar UK, Dayal R, Chaturvedi R, Kumar R. 2021. Freshwater fish diversity database of Central India: Implementation and utility. Indian Journal of Fisheries, 68(2).

Qadri S. 2022. Fish Fauna of Kashmir Valley and Their Conservational Measures for Sustainable Fish Production. In Biodiversity, Conservation and Sustainability in Asia: Volume 2: Prospects and Challenges in South and Middle Asia (pp. 521-528). Cham: Springer International Publishing.

Qayoom I, Akhtar M, Dar SA, Khan S, Hussain N, Bhat BA, 2019. A study of socio-economic status of fisher communities in district Baramulla of Jammu & Kashmir. J. Pharmacogn. Phytochem 8, 583–586.

Rahim KA, Khalijan DS, Esa Y, Rena IE. 2009. Freshwater fish diversity and composition in Batang Kerung Flood plain, Balai Ringin, Sarawak. Pertanika J Tropica Agri Sci 31:7–16

Raju CS, Rao JCS, Rao KG, Simhachalam G. 2016. Fishing methods, use of indigenous knowledge and traditional practices in fisheries management of Lake Kolleru. Journal of Entomology and Zoology Studies 4(5):37-44.

Rathore, V. and Dutta, S. P. S. (2015). Fish fauna of river Ujh, an important tributary of the river Ravi, District Kathua, Jammu. Environment Conservation Journal, 16(1&2), 81-86.

Romshoo SA, Rashid I, Altaf S, Dar GH. 2020 Jammu and Kashmir State: an overview. Biodiversity of the Himalaya: Jammu and Kashmir State pp.129-166.

Sarkar U, Gupta B, Lakra W. 2010. Biodiversity, ecohydrology, threat status and conservation priority of the freshwater fishes of river Gomti, a tributary of river Ganga (India). Environmentalist 30:3– 17.

Sarkar UK, Pathak AK Sinha RK, Sivakumar K, Pandian AK, Pandey A, Dubey VK, Lakra WS. 2012. Freshwater fish biodiversity in the River Ganga (India): Changing pattern, threats and conservation perspectives. Rev Fish Biol Fish 22: 251-272

Sarma D, Akhtar MS, Singh AK. 2018. Coldwater fisheries research and development in India. Aquaculture in India pp.93-133.

Sehgal KL. 1999. Coldwater fish and fisheries in the Indian Himalayas: rivers and streams. p. 41-63. In T. Petr (ed.) Fish and fisheries at higher altitudes: Asia. FAO Fisheries Technical Paper 385. FAO, Rome. 304 p.

Sharma S, Sharma P, Siddique A, Belsare DK. 2007. Fish species diversity of Sirpur lake, Indore (M.P.), India. Him J Environ Zool 18:349–352

Singh G. Agarwal NK. 2014. Fishing methods in upper Ganga River system of central Himalaya, India. Journal of Fisheries 2(3): 195-202.

Sinha, M., Khan, M.A. and Jha, B.C. (1999). Ecology, Fisheries and Fish Stock Assessment of Indian Rivers. CIFRI Bulletin (90).

Sivakumar K. 2008. Species richness, distribution pattern and habitat use of fishes in the Trans-Himalayas, India. Electronic Journal of Ichthyology 1: 31-42.

Syed N, Mohite A, Sadawarte R, Desai A, Shah TH. 2020. Design aspects of fishing crafts and gears of Wular Lake of Kashmir, India. Journal of Experimental Zoology India 23.

Talwar PK, Jhingran AG. 1991. Inland fishes of India and adjacent countries (Vol. 2). CRC press.

Tilak R. 1990. On the fish fauna of Ladakh with notes on Zoogeography. Records of the Zoological Survey of India 131-149.

Vass KK, Gopakumar K. 2002. Coldwater fisheries and research status in India. Highland fisheries and aquatic resource management (eds. Vass KK and Raina HS) NRCCWF, Bhimtal, India, 3-29.

Wiley MJ, Kohler SL, Seelbach PW. 1997. Reconciling landscape and local views of aquatic communities: lessons from Michigan trout streams. Freshw Biol 37:133–148.

Yousuf Z, Tasaduq HS, Farooz AB, Adnan A, Bilal AB, Irfan A, Shah FA, and Nimat S. 2023. Fishing gears and craft of Anchar Lake: design specifications and operational aspects. Journal of Experimental Zoology India 26: 2.

5

Biology and Ecology of Schizothoracinae from Kashmir Waters

Ifrah Rashid*, Tasaduq H. Shah, Ahali Jahan, Sobiya Gul and Zarka Yousuf

Division of Fisheries Resource Management, Faculty of Fisheries, SKUAST-K Rangil Ganderbal, Kashmir, Jammu and Kashmir

Abstract

Snowtrouts, resilient coldwater fish species of the Cyprinidae family and Schizothoracinae subfamily, are found in the pristine waters of the Kashmir Himalayan and Sub-Himalayan regions. The genus Schizothorax, which includes 63 species worldwide, has 28 species native to the Indo-Himalayan region, with only five species indigenous to Kashmiri waters. These coldwater fish primarily consume algae attached to rocks and typically spawn once a year. The decline of snow trout species in the Himalayan region is a complex issue with multiple immediate causes. It is crucial to develop comprehensive strategies to protect these iconic fishes at this critical juncture. Efforts must focus on safeguarding and restoring their natural habitats and creating innovative conservation plans to prevent further population declines and conserve these valuable fish resources. This chapter provides an in-depth overview of the ecology and biology of Schizothoracinae in the waters of Kashmir, emphasizing their adaptations, habitat preferences, feeding habits, reproductive biology, and ecological significance.

Keywords: *Schizothorax, Snowtrouts, Algae, Coldwater, Kashmir, Himalayas*

Introduction

Schizothoracinae, commonly known as snow trouts, are a specialized group of fish thriving in the fast-flowing, snow-fed streams and lakes of regions such as Afghanistan, Pakistan, the Himalayas in India and Nepal, China, and parts of the former USSR. Adapted to these challenging environments, Schizothoracinae have not been extensively studied due to the difficult terrain they inhabit. In India, they are primarily found along the Himalayan range, with an exception in the Periyar, Kerela (river and lake). The Indian Schizothoracinae are classified into seven genera, forming a significant part of the Himalayan fishery. Resembling Salmonidae, they are commonly referred to as "Indian Trout." Early mentions of *Schizothorax* species by Day (1877, 1889) were brief. Subsequent taxonomic studies by researchers such as Hora (1921, 1934, 1937, 1953), Misra (1949), and Mukerji (1936) have provided a more comprehensive understanding of their taxonomy.

The Schizothoracinae, part of the Cyprinidae family, represent a diverse group uniquely adapted to high-altitude environments and have been categorized into three groups based on their morphological characteristics: "primitive group," "specialized group," and "highly specialized group" (Cao *et al.*, 1981). The primitive group includes *Schizothorax* and *Aspiorhynchus*, which share more traits with the outgroup *Barbodes hexagonolepis*. They are distinguished by the presence of uroneuralia, three or four rows of pharyngeal teeth, minimal scale degeneration, and a lack of pronounced sexual dimorphism. The specialized group includes *Ptychobarbus, Gymnodiptychus,* and *Diptychus*, and is marked by the absence of uroneuralia, one or two rows of pharyngeal teeth, and moderate scale degeneration. The highly specialized group, which comprises *Gymnocypris, Oxygymnocypris, Schizophygopsis, Chuanchia,* and *Platypharodon,* is characterized by the complete absence of barbels and scales and a well-developed canalis preoperculomandibularis (Qi *et al.*, 2012, Yonezawa *et al.*, 2014). The genus *Schizothorax* comprises 63 species in total. Among these, 34 species are mainly found in Central Asia, while 28 species inhabit the Indo-Himalayan region. The species of Schizothorax in the Himalayan region include *S. curvifrons, S. nasus, S. richardsonii, O. sinuatus, S. planifrons, S. esocinus, Schizothoraichthys progastus, S. longipinnis, S. kumanonensis, O. moleswortbii, S. hugelli, S. labiatus,* and *S. micropogon* (Menon, 1999; Mishra, 2003; Zhang *et al.*, 2018; Ma *et al.*, 2020).

Hora (1955) and Menon (1962) researched the evolution of schizothoracines and determined that they first emerged during the initial interglacial period, a time when the formation of turbulent streams in Central Asia prompted a reduction in scales, a characteristic feature of schizothoracines. Presently, primitive forms of this group can be found in South China. Favorable environmental conditions during the second glacial period allowed them to migrate westward, reaching regions such as Kashmir and Sistan. The extensive diversification of genera and species within the sub-family Schizothoracinae likely took place during the second and subsequent interglacial periods. Today, schizothoracines are predominantly found in Central Asia, with some species inhabiting the southern slopes of the Himalayas. Their distribution is likely attributed to the Trans-Himalayan origins of significant rivers like Indus, Sutlej, and Brahmaputra, which enabled their descent to the lower reaches of these rivers (Sehgal, 1999).

The Union Territory of Jammu and Kashmir is rich in rivers, lakes, and reservoirs. The Himalayas in this region feature numerous watersheds, influencing the distribution of fish species by environmental factors such as water current velocity, substrate nature, and food availability. The water bodies exhibit significant seasonal variations in their physical, chemical, and biological characteristics, supporting diverse aquatic life. The cold, clear waters provide a habitat for a unique array of fish fauna (Bhat *et al.*, 2020). The group was first described by Heckel in 1838 from a collection by Von Huegel in Kashmir, identifying ten different species. McClelland added three more species in 1839 and several more in 1842, recognizing their distinctiveness from other Cyprinidae and proposing the sub-family name Schizothoracinae. These fish are characterized by minute scales covering their bodies and a naked membranous area in front of the anal fin, flanked by vertically placed, enlarged scales resembling eave tiles, adaptations suited to their hill-stream environments. Schizothoracinae exhibit numerous adaptations for survival in the fast-flowing mountain streams and plateau lakes of Central Asia, the Himalayas, and the Tibetan Plateau. These fish are economically significant, especially in the Kashmir Valley and other Himalayan regions, where they are prevalent in local fish markets.

Schizothoracines are believed to have migrated to Kashmir's lakes and streams from central Asian watersheds near the Suleiman Ranges, the inner regions of the North Western Himalayas, and the southern slopes of the Hindukush and Karakorum mountains. Tectonic activities isolated these fish in the Kashmir region, leading to the evolution of numerous endemic species (Talwar and Jinghran, 1991; Yousuf, 1996; Bhat *et al.*, 2010). In the challenging mountainous terrain, Schizothoracinae experienced prolonged evolution, resulting in several adaptative mechanisms. These adaptations include a reduction in the number of scales, pharyngeal teeth, and barbels, as well as changes in body shape to more rounded and depressed forms (Talwar and Jinghran, 1991). However, their populations are declining in many of Kashmir's water bodies, including Dal Lake, Jhelum River, Sindh River, Manasbal Lake, and Wular Lake, due to pollution, habitat degradation, introduction of carp, and sand and gravel extraction (Balkhi, 2007).

In the Kashmir valley, 14 species of *Schizothorax* were previously identified, out of which only 5 species (*S. curvifrons, S. esocinus, S. labiatus, S. plagiostomus*, and *Schizopyge niger)* are currently reported (Jan and Ahmed, 2020). The taxonomy of *Schizothorax* has been studied by numerous researchers (Heckel, 1838; McClelland, 1842; Hora 1936, Silas, 1960 Talwar, and Jhingran, 1991, Misra, 2003).Despite these studies, there is still debate regarding the precise number of species within this genus due to overlapping morphological traits. This uncertainty is particularly pronounced in the various aquatic habitats of Kashmir valley, complicating the study of this species complex. Additionally, the hybrids of some species have further muddied their taxonomic and evolutionary understanding (Hora, 1936).

Key Schizothorax species in Kashmir

1. *Schizothorax plagiostomus* Heckel, 1838

Classification

Order	:	Cypriniformes
Family	:	Cyprinidae
Subfamily	:	Schizothoracinae
Genus	:	Schizothorax
Species	:	*S. plagiostomus*

Local name: Khont

Characteristics: *S. plagiostomus* exhibits an elongated, fusiform body with a distinctively projecting snout. It has a distint inferior and wide mouth, characterized by a deep and short lower jaw with a sharp, keratinized cutting edge at the front. The lower lip fold is expanded and papillose. The pharyngeal teeth are organised in three rows. Specifically 5, 3, 2/2, 3, 5 (2). The first tooth in the main row is considerably smaller than the others, conical, and recurved, while the second tooth is the largest, somewhat swollen at the apex with a recurved tip. All other teeth have a semitruncate tip. A series of enlarged scales runs along the base of anal fin. The scales are quite small, numbering between 89 and 99 along the lateral line. The dorsal fin, inserted opposite to pelvic fins, has 7-8 branched rays, while the anal fin has 5 branched rays. The caudal fin is deeply emarginated. Additionally, the gill rakers are 15 (1), 17 (5), 18(4), 19(3), 20(6), 21(4), 22(1) (Kullander *et al.*, 1999).

Distribution: Geographically, *S. plagiostomus* inhabits various rivers and tributaries across the Himalayas, extending into China, Afghanistan, Pakistan,

Turkistan, Nepal, Ladakh, Tibet, Bhutan, and northeastern India. In India, it is a significant food fish in the Himalayan region, particularly in Kashmir, Himachal Pradesh, Uttarakhand, the Uttar Pradesh foothills, and Assam (Day, 1958).

Conservation Status: Vulnerable (IUCN, 2023)

2. *Schizothorax labiatus* (McClelland, 1842)

Classification

Order	: Cypriniformes
Family	: Cyprinidae
Subfamily	: Schizothoracinae
Genus	: Schizothorax
Species	: *S. labiatus*

Local name: Chush

Characteristics: The body shape is elongated and fusiform, featuring an upper jaw that protrudes forward and a lower jaw characterized by wide lip folds, typically separated by a raised pad. The pharyngeal teeth are arranged in three rows: 5, 3, 2/2, 3, 5 (2). The first tooth in the primary row is notably smaller and conical, curving backward, while the second tooth is the largest, slightly swollen at the tip, and also curved backward. The remaining teeth all have curved or semi-truncated tips. Enlarged scales run along the base of the anal fin, and the lateral line consists of very small scales, numbering between 85 and 104. The branched dorsal and anal fin rays are 7(13), 18(12) and 5 (25) respectively. Additionally, the gill rakers are 14 (1), 15 (6), 16(7), 17(5), 18(5), 19(1) (Kullander *et al.*, 1999).

Distribution: Geographically, *S. labiatus* is found in various rivers and tributaries across the Afghanistan, Pakistan, Nepal, India (Talwar and Jhingran, 1991) and Tibet (Yunfei, 1987).

Conservation Status: Least concern (IUCN, 2023)

3. *Schizothorax curvifrons* Heckel, 1838

Classification

Order : Cypriniformes

Family : Cyprinidae

Subfamily : Schizothoracinae

Genus : Schizothorax

Species : *S. curvifrons*

Local name: Satter gad

Characteristics: The body is elongated and fusiform, featuring a short, blunt, and slightly protruding upper jaw. Mouth crescent shaped, broader than long, lower jaw without horny cover. The lips are thin and not expanded into wide folds. Barbles much smaller than eyes. Along the base of the anal fin, there is a series of enlarged scales. Pharyngeal teeth are arranged in three rows: 5, 3, 2-1-2, 3, 5 (2). The first tooth in the main row is much smaller and conical, curving backward, while the second tooth is the largest, somewhat swollen at the tip, and truncated in two aspects. The remaining teeth all have truncated tips, forming a semi-dorsal aspect. The lateral line consists of very small scales, numbering between 95 and 121. The origin of dorsal fin is opposite the ventrals and nearly midway between the tip of snout and base of the caudal fin. Height of the body is somewhat more than the length of head and ¼ of the total length. The branched dorsal and anal fin rays are 6(10), 7(15) and 5 (25) respectively. Additionally, the gill rakers are 21 (1), 22 (1), 23(1), 24(6), 25(5), 26(5), 27 (3), 28(3) (Talwar and Jinghran, 1991; Kullander *et al.*, 1999).

Distribution: Geographically, *S. curvifrons* is found in various lakes, rivers, and tributaries across the Afghanistan, Pakistan, India (Menon, 1999) and China (Walker and Yang, 1999). It has also been reported in Iran, Uzbekistan,

Kazakhstan and Kyrgyzstan (Berg, 1964; Coad, 1995; Kamilov and Urchinov, 1995).

Conservation Status: Least concern (IUCN, 2023)

4. *Schizopyge niger* (Heckel, 1838)

Classification

Order	:	Cypriniformes
Family	:	Cyprinidae
Subfamily	:	Schizothoracinae
Genus	:	Schizothorax
Species	:	*S. niger*

Local name: Ael gad

Characteristics: The body is clongated and fusiform, with a short, blunt, and slightly protruding upper jaw. Mouth inferior, horse shoe shaped nearly horizontal, broader than long, margin of lower jaw cutting, its inside covered with deciduous cartilaginous layer, lower labial fold interrupted in the middle, barbels shorter than eyes. The lips are thick but not expanded into wide folds. Along the base of the anal fin, there is a series of enlarged scales. Pharyngeal teeth are arranged in three rows: 5, 3, 2/2, 3, 5 (2). The first tooth in the main row is much smaller and conical, pointed rather than curved; the second tooth is the largest, somewhat swollen at the tip, and semi-recurved; all other teeth have a semi-recurved tip. The lateral line consists of very small scales, numbering between 92 and 100. The origin of dorsal fin a liitle nearer to the root of caudal fin than to the end of snout. The branched dorsal and anal fin rays are 6(9), 7(1) and 5 (10) respectively. Additionally, the gill rakers are 12 (1), 13 (2), 14(4), 16(2), 18(1) (Talwar and Jhinghran, 1991; Kullander *et al.*, 1999).

Distribution: It is found in cold lakes and nearby channels in Kashmir region in India and Pakistan (Talwar and Jhingran, 1991)

Conservation Status: Not evaluated (IUCN, 2023)

5. *Schizothorax esocinus* Heckel, 1838

Classification

Order : Cypriniformes

Family : Cyprinidae

Subfamily : Schizothoracinae

Genus : Schizothorax

Species : *S. esocinus*

Local name: Chhirruh

Characteristics: The fish is elongated and fusiform, possessing a lengthy snout, with a slightly protruding upper jaw or jaws of equal length. Head depressed, margins of the lower jaw not cutting, without cartilaginous covering. Barbles longer than eyes. The Pharyngeal teeth are arranged in three rows in a pattern of 5, 3, 2 / 2, 3, 5 (2). The first tooth in the primary row is notably smaller and conical, with a recurved shape; the second tooth is the largest, somewhat swollen and curved at an apex, while all other teeth have recurved tips and expanded apical regions, bent forward to expose the grinding surface dorsally. Additionally, it features a row of enlarged scales along the base of its anal fin. The scales are generally small, numbering between 96 and 108 along the lateral line. Origin of the dorsal fin is somewhat nearer to root of caudal than to end of snout. Body covered with numerous black dots. The branched dorsal and anal fin rays are 7(3), 8(22) and 5 (25) respectively. Additionally, the gill rakers are 8(1), 9 (2), 10(3), 11(6), 12(8), 13(3), 14(1), 15(1) (Kullander *et al.*, 1999, Talwar and Jhingran, 1991).

Distribution: It is found in mountain streams, rivers and lakes of Afghanistan, Pakistan, India. The fish is also reported from Nepal and China (Shrestha, 1990; Wu and Wu, 1992).

Conservation Status: Vulnerable (IUCN, 2023)

Morphological Adaptations and Habitat Preferences

Schizothoracine fish exhibit remarkable morphological adaptations to thrive in their cold, fast-flowing aquatic environments. Their streamlined bodies, elongated fins, and reduced or scaleless skin reduce drag and enable efficient movement in turbulent waters (Talwar and Jhinghran, 1991; Mushtaq, 2024). Additionally, these fish possess physiological adaptations such as well-developed gill structures and high hemoglobin concentrations, which enhance oxygen uptake in oxygen-deficient conditions (Bigham *et al.*, 2010; Wang *et al.*, 2022).

Habitat Preferences: Schizothoracine fish are primarily found in clear, cold-water habitats with rocky substrates, often in glacier-fed rivers, streams, and lakes. They show preferences for specific microhabitats within these ecosystems: shallow riffles with moderate to fast currents are favored for feeding and spawning, while deeper pools provide refuge during periods of high flow (Talwar and Jhinghran, 1991).

Schizopyge niger: A benthopelagic, potamodromous species inhabiting freshwater lakes and adjoining channels (Riede, 2004).

Schizothorax esocinus: A benthopelagic, potamodromous species inhabiting the mountain streams, rivers, and lakes, usually residing on sandy and gravel substrates (Shrestha, 1990; Menon, 1999).

Schizothorax curvifrons: A benthopelagic, potamodromous species occurring in freshwater rivers, swamps, and lakes, also favoring sandy and gravel substrates (Raina and Petr, 1999; Riede, 2004).

Schizothorax plagiostomus: A benthopelagic, potamodromous species inhabiting freshwater rivers and streams (Riede, 2004).

Schizothorax labiatus: A benthopelagic species found in high-altitude rivers and hill streams (Baensch and Riehl, 1997).

Feeding Habitats of Schizothoracids

Schizothoracine fish are integral to nutrient cycling and energy flow in their ecosystems due to their diverse feeding habits and ecological interactions. They are opportunistic feeders, consuming a wide array of food resources. Primarily, they feed on benthic organisms such as aquatic insects, larvae, crustaceans, and mollusks, scraping these off rocks and sediments with specialized feeding structures. Their diet also includes algae, detritus, and occasionally terrestrial insects that fall into the water.

S. plagiostomus: Consumes detritus (Raina and Petr, 1999).

S. labiatus: Feeds on a mixed diet of plants and animals (Raina and Petr, 1999).

S. curvifrons: Diet consists of both animals and plants (Berg, 1964).

S. esocinus: Primarily herbivorous, feeding on bottom detritus (Raina and Petr, 1999).

S. niger: Primarily herbivorous, feeding on bottom detritus (Raina and Petr, 1999)

Numerous studies have detailed the feeding habits and habitats of schizothoracine fish in the Kashmir valley, as summarized in the (Table 1).

Table 1. Detailed the feeding habits and habitats of schizothoracine fish in the Kashmir valley.

Species	Food and feeding habitat	Reference
S. niger	Herbivore; feeding on plant and animal matter (67.53% and 23.43% respectively). Green algae and crustaceans being the major contributor to plant and animal food respectively.	Qadir *et al.*, 2024
	Herbivore; feeding mainly on green algae, plant fragments, diatoms, detritus, unrecognizable matter	Sabha *et al.*, 2017
	Detritophagus and herbivorous	Sunder *et al.*, 1979
	Surface feeder; herbivore; plant and animal food (63.5% and 33.5% respectively)	Jan and Das, 1970
S. esocinus	Omnivorous; feeding on diatoms, macrophytes and crustaceans	Hafiz *et al.*, 2023
	Feeds on diatoms (39.43%) followed by crustaceans (26%)	Mir, 2019
	Typical herbivore; plant and animal matter (63.5% and 30.5% respectively)	Jan and Das, 1970
S. curvifrons	Illiophagic and herbivore; occasional column feeder; feeding on decayed organic matter (40.33%), sand and mud (17.51%), phytoplankton (38.78%), zooplankton (2%) and miscellaneous matter (1.38%)	Sunder (1984) and Sunder and Subhla (1985)
S. labiatus	Omnivore; Feeds on plant matter (67.42%), animal matter (23.67%), sand and mud (6.52%), miscellaneous matter (2.41%).	Farooq, 2024
	Feeds on algae, plant fragments, detritus, sand/silt and zooplankton; detritus being dominant (50%), followed by algae (26.7%), sand/silt (18.5%), plant matter (4.7%) and zooplankton (0.1%)	Arafat, 2020

S. plagiostomus	Benthic herbivore; feed mainly on plant matter (62.02%), miscellaneous food (mud, sand, detritus) being 31.01% and a small quantity of animal food (6.97%)	Jan *et al.*, 2018
	Herbivore; are essentially fed on diatoms. Other food items included weeds, decayed organic matter and green algae	Sunder and Bhagat, 1979
	Bottom-feeding Herbivore; feeding on fine plant matter (67%), large ostracods, copepods, insects (24%) and detritus	Jan and Das, 1970

Reproductive Biology of Schizothoracids

The reproductive biology of schizothoracine fish is closely aligned with seasonal changes in their high-altitude habitats, featuring unique strategies to cope with these challenging environments. Most schizothoracine species undergo seasonal reproductive migrations, traveling long distances to reach spawning grounds in shallow, well-oxygenated riffles and tributaries during the spring and summer. These migrations and spawning behaviors are often triggered by environmental cues such as water temperature, flow, and photoperiod (Talwar and Jhinghran, 1991).

Spawning Behavior and Habitat

General Spawning Behavior: *Schizothorax* species, with the exception of *S. niger*, migrate from lakes and rivers in the valley to nearby tributaries for spawning (Talwar and Jhinghran, 1991). Female schizothoracines lay eggs in shallow stream areas with sandy, gravelly, or pebbly bottoms, often covered with debris, in moderate currents (Talwar and Jhinghran, 1991). Adults spawn in the gravel and sandy beds of incoming streams, whereas fry are generally located in quieter sections of streams or side branches. *S. niger* typically breeds near the roots of willow trees in the cold and clear regions of the lakes (Raina and Petr, 1999). Spawning periodicity of *Schizothorax* species are presented in (Table 2).

Table 2. *Schizothorax* species spawning periodicity

Species	Spawning periodicity
S. niger	Annual spawner; spawns once from Feb-March
S. esocinus	Annual spawner; spawns once from March- July
S. curvifrons	Annual spawner; spawns once from May-August
S. labiatus	Annual spawner; spawns once from April-May
S. plagiostomus	Spawns twice; September-October; March-April

Life History Strategies

Schizothoracine fish show a range of life history strategies suited to their cold, nutrient-poor environments. These strategies include slow growth rates, delayed maturity, and low reproductive outputs, traits typical of species living in extreme conditions (Chen and Cao, 2000). Juveniles often inhabit shallow, nutrient-rich nursery habitats and gradually transition to the deeper, faster-flowing waters as they mature. Numerous studies have investigated the reproductive biology of schizothoracine fish in the Kashmir region, providing detailed insights into their complex life cycles and environmental adaptations. Length range and fecundity of *Schizothorax* species presented in (Table 3).

Table 3. *Schizothorax* species length range and their fecundity.

Species	Length range (mm)	Fecundity	Reference
S. niger	195-252	1,802-5,508	Malhotra, 1965
	135-365	810-11,458	Jyoti and Malhotra, 1972
	170-280	5,618-17,367	Sunder *et al.*, 1979
	215-331	5471-18385	Hussain *et al.*, 2018
	132-288	3,359-15,712	Ali *et al.*, 2020
	195-280	2,700-16,758	Wani *et al.*, 2022
S. esocinus	140-260	1,687-11,470	Rampal, 1967
	195-485	3,910-53,106	Raina, 1977
		4,573-19,976	Sunder and Bhagat, 1979
S. curvifrons	196-411	4,843-31,221	Sunder and Subla, 1984
	200-325	4,762-17,196	Qadri *et al.*, 2015
S. labiatus	143-350	3,857-16,472	Farooq, 2015
	266-324	2,754-13,465	Arafat and Bakhtiyar, 2022
S. plagiostomus	216-331	2,600-16,605	Das and Kaul, 1965
	240-300	6,700-27,500	Rampal, 1967
	241-331	2,598-27,312	Mir, 1979
	345-540	3,474-13,916	Agarwal *et al.*, 1988
	262-453	3,437-34,800	Jan *et al.*, 2014

Population Dynamics and Ecological Significance

Understanding the population dynamics and ecological significance of schizothoracine fish is crucial for their safeguarding and governance amid rising anthropogenic pressures and environmental changes.

Population dynamics: Schizothoracine fish populations are influenced by an intricate interaction of biotic and abiotic factors, such as predation, habitat availability, competition, water flow, and temperature regimes. Their populations exhibit fluctuations in abundance and distribution across seasons and years, reflecting the dynamic nature of their habitats.

Ecological significance: Schizothoracine fish are keystone species in high-altitude freshwater ecosystems, playing critical roles in maintaining ecosystem structure and function. They contribute to biodiversity, nutrient cycling, and energy flow, thereby influencing the overall health and resilience of their habitats. Additionally, schizothoracine fish serve as important bioindicators of environmental quality, reflecting changes in water chemistry, habitat integrity, and ecosystem productivity. Their presence or absence can signal shifts in ecological processes and serve as early warning signs of environmental degradation.

Conservation concerns: Despite their ecological importance, schizothoracine fish face numerous threats to their survival, primarily driven by human activities and climate change. Habitat degradation, including dam construction, deforestation, and pollution from agricultural runoff and urbanization, poses significant challenges to their long-term viability (Petr, 1999). Overfishing, both for subsistence and commercial purposes, further exacerbates population declines, particularly in vulnerable species with restricted distributions.

Conservation strategies: Efforts to conserve schizothoracine fish and their habitats require a multifaceted approach, integrating scientific research, community engagement, and policy interventions. Key conservation strategies include the establishment of protected areas and freshwater reserves, implementation of sustainable fishing practices, habitat restoration initiatives, and public education and awareness campaigns. Effective conservation planning and management requires collaborative partnerships among government agencies, non-governmental organizations, local communities, and scientific institutions.

Future directions: Continued research into the ecology, biology, and conservation of schizothoracine fish is essential for addressing knowledge gaps and informing evidence-based management decisions. Long-term monitoring programs, genetic studies, and population assessments can provide valuable insights into population trends, genetic diversity, and adaptive potential in response to environmental change. Additionally, interdisciplinary approaches that integrate traditional ecological knowledge with modern scientific methodologies can enhance our understanding of human-fish interactions and support sustainable resource management practices in mountainous regions.

Conclusion

Schizothoracine fish represent a unique and invaluable component of high-altitude freshwater ecosystems, embodying the resilience and adaptability of organisms to extreme environmental conditions. Their intricate ecological interactions, morphological adaptations, and ecological significance underscore the importance of conserving these iconic species and their habitats for future generations. Through collaborative efforts and proactive conservation measures, we can ensure the continued survival and ecological integrity of schizothoracine fish populations in the face of ongoing environmental challenges.

References

Agarwal, N.K., Singh, Wazir. and Singh, H.R., 1988. Fecundity of the snow-trout, Schizothorax plagiostomus (Heckel) from Garhwal Himalaya.

Ali, S., Shah, T.H., Bhat, A. and Rashid, I., 2020. Studies on the reproductive potential of Schizopyge niger (Heckel, 1838) from Kashmir. Journal of Pharmacognosy and Phytochemistry, 9(4S), pp.613-616.

Arafat, M. Y. 2020. Ichthyofauna of Vishav Stream with special reference to bioecology of Schizothorax labiatus McClelland 1842. PhD thesis, University of Kashmir.

Arafat, M.Y. and Bakhtiyar, Y., 2022. Reproductive pattern and maturity phases of indigenous Kunar snowtrout Schizothorax labiatus inhabiting Vishav Stream in Kashmir Himalaya, India. Environmental Biology of Fishes, 105(2), pp.247-260.

Baensch, H.A. and R. Riehl, 1997. Aquarien Atlas, Band 5. Mergus Verlag, Melle, Germany. 1148 p.

Balkhi, M. H. 2007. Fish diversity in Jammu and Kashmir and conservation measures. In: Kashmir Speaks (Ed Riyaz A. Patloo) 6: 104-115.

Berg, L.S., 1964. Freshwater fishes of the U.S.S.R. and adjacent countries. Volume 2, 4th edition. Israel Program for Scientific Translations Ltd, Jerusalem. (Russian version published 1949).

Bhat, F. A., Balkhi, M. H. and Yousuf, A. R. 2010. Fish diversity in the Kashmir Himalaya. In: Biodiversity, development and Poverty elevation; International Day for Biological Biodiversity. P. G. Deptt. of Botany, University of Kashmir. pp. 24-27.

Bhat, F.A., Yousuf, A.R. and Balkhi, M.H., 2020. Diversity of fishes in Jammu and Kashmir State. Biodiversity of the Himalaya: Jammu and Kashmir State, pp.859-887.

Bigham, A., Bauchet, M., Pinto, D., Mao, X., Akey, J.M., Mei, R., Scherer, S.W., Julian, C.G., Wilson, M.J., López Herráez, D. and Brutsaert, T. 2010. Identifying signatures of natural selection in Tibetan and Andean populations using dense genome scan data. PLoS genetics, 6(9):p.e1001116.

Cao, W.X., Chen, Y.Y., Wu, Y.F. and Zhu, S.Q., 1981. Studies on the period, amplitude and type of the uplift of the Qinghai-Xizang Plateau.

Chen, Y. and W. Cao, 2000. Cyprinidae: Schizothoracinae. p. 273-390. In P. Yue, et al. (Eds). Fauna Sinica. Osteichthyes. Cypriniformes III. Science Press. Beijing. 1-661.

Coad, B.W., 1995. Freshwater fishes of Iran. Acta Sci. Nat. Acad. Sci. Brno. 29(1):1-64.

Das, S.M. & B.L. Koul. 1965 The gross anatomy of gonads and fecundity of four fishes of Kashmir. Kash. Sci. 2(1-2): 64-76.

Day, F., 1877. Geographical Distribution of Indian Freshwater Fishes.—Part I. The Acanthopterygii, Spiny-rayed Teleostean Fishes. Zoological Journal of the Linnean Society, 13(67), pp.138-155.

Day, F., 1889. Fauna of British India, including Ceylon and Burma. Flshes, 1, XVII+548 pp. 64 figs.-London (Taylor & Francis).

Day, F., 1958. Fishes of India, Volumes 1-2. William Dawson and Sons Ltd., London.

Farooq, I. 2015. Study on the Growth and Reproductive biology of Schizothorax labiatus McClelland in River Jhelum, Kashmir. MFSc Thesis, SKUAST-K

Farooq, I. 2024. Biology and life history traits of snowtrout Schizothorax labiatus (McClelland) in river Jhelum, Kashmir. PhD thesis, SKUAST-K.

Hafiz, Z., Bhat, F.A., Shah, T.H., Abubakr, A., Kumar, A., Bhat, B.A., Altaf, M. and Bazaz, A.I. 2023. Study on gut morphometric parameters of Schizothorax esocinus Heckel, 1838. Gut, 1, p.3

Heckel, J. J. 1838. Fishes from Cashmere. Collected and edited by Carl Freiherren, von Hugel etc. Vienna, I-X + 1-12 (In German).

Hora, S., 1955. Fish and fisheries in ancient India. Trans. Bose Res. Inst., Calcutta, 20, pp.145-150.

Hora, S.L., 1921. Fish and fisheries of Manipur with some observations on those of the Naga Hills. Records of the Zoological Survey of India, pp.165-214.

Hora, S.L., 1934. Notes on fishes in the Indian Museum. XXIII. On a collection of fish from the S. Shan States, Burma. Records of the Zoological Survey of India, pp.353-370.

Hora, S.L., 1936. On a further collection of fish from the Naga Hills. Records of the Zoological Survey of India, pp.317-331.

Hora, S.L., 1937. Comparison of the fish-faunas of the northern and the southern faces of the great Himalayan range. Records of the Zoological Survey of India, pp.241-250.

Hora, S.L., 1953. Mollusc control through fish farming. Mollusc control through fish farming.

Hussain, S., Bhat, F.A., Maqsood, H.M., Balkhi, M.H., Majid, I. and Najar, A.M. 2018. Present status of breeding biology of Schizothorax niger in Dal Lake Kashmir. Journal of Entomology and Zoology Studies, 6(6): 930-935.

IUCN, 2023. The IUCN Red List of Threatened Species. Version 2023-1. .

Jan, K. and Ahmed, I. 2020. Morphometric and meristic characters of snow trout, inhabiting the Jhelum River and its tributaries. Fisheries & Aquatic Life, 28(4):216-224.

Jan, M., Jan, N. and Khan, I.A. 2018. Food and feeding habits of snow trout, Schizothorax plagiostomus in River Lidder, from Kashmir Himalaya. Bulletin of Pure & Applied Sciences-Zoology, 37(1): 1-10.

Jan, M., Jan, U. and Shah, G.M. 2014. Studies on fecundity and Gonadosomatic index of Schizothorax plagiostomus (Cypriniformes: Cyprinidae). Journal of Threatened Taxa, 6(1): 5375-5379.

Jan, N. A. and Das, S. M. 1970. Qualitative and quantitative studies on the food of eight fishes of Kashmir valley. Ichthyologica, 10(1-2): 20-26.

Jyoti, M. K. and Malhotra, Y. R. 1972. Studies on the fecundity of Schizothorax niger Heckel from Dal Lake (Kashmir). Indian J. Expt. Biol, 10 (1): 74-76.

Kamilov, G. and Z.U. Urchinov, 1995. Fish and fisheries in Uzbekistan under the impact of irrigated agriculture. p. 10-41 In T. Petr (ed.) Inland fisheries under the impact of irrigated agriculture: Central Asia. FAO Fisheries Circular No. 894.

Kullander, S.O., Fang. F., Delling, B. and Ahlander, E. 1999. The fishes of the Kashmir Valley. In: L. Nyman (ed) River Jhelum, Kashmir Valley, Impact on the aquatic environment. SWEDMAR, Sweden. pp. 100-167.

Ma, B., Zhao, T., Xu, B., Zhong, L., Wu, X., Wei, K., Zhang, Z. and Li, Y., 2024. Morphological variation in Schizothorax oconnori, Schizothorax waltoni (Teleostei: Cyprinidae: Schizothoracinae), and their natural hybrids from the middle Yarlung Zangbo River, Tibet. Ecology and Evolution, 14(5), p.e11342.

Malhotra, Y .R. 1965. Studies on the ovarian cycle of Kashmir fishes Schizothorax niger Heckel and Botia birdi Chaudhary with observations on nucleolar extrusions and problems of corpus leutem in fishes. Ph.D. thesis, J & K Univ.

McClelland, J. 1839. Indian cyprinidae. Asiat. Res. Calcutta, 19(2): 262-450.

Mcclelland, J. 1842. Afghan collection of fishes. Calcutta J. nat. Hist. Calcutta, 2, pp. 575-589.

Menon, A.G.K., 1962. A distributional list of fishes of the Himalayas. J. Zool. Soc. India, 14(1), pp.23-32.

Menon, A.G.K., 1999 Check list - fresh water fishes of India. Rec. Zool. Surv. India, Misc. Publ., Occas. Pap. No. 175, 366 p.

Mir, S. 1979. Studies on the biology of Oreinus plagiostomus (Mc clleland). Ph.D. thesis, University of Kashmir

Mir, S. 2019. Studies on the food and growth of Schizothorax esocinus Heckel, 1838 in river Jhelum, Kashmir. MFSc Thesis, SKUAST K.

Mishra, K. S. 2003. An aid to commercial fisheries. Narendra publishing house India

Misra, K.S. 1949. A note on the systematic position of two cyprinid fishes of the genus Oreinus McClelland from Kashmir. J. Zool. Soc. India, 1(1): 39-40.

Mukerji, S.K., 1936. Contributions to the autecology of Mercurialis perennis L. Journal of Ecology, 24(1), pp.38-81.

Petr, T., 1999. Coldwater fish and fisheries in Afghanistan. p. 138-148. In T. Petr (ed.) Fish and fisheries at higher altitudes: Asia. FAO Fish. Tech. Pap. No. 385. FAO, Rome. 304 p.

Qadir, G., Sidhu, H.K. and Ahmed, I., 2024. Food and Feeding Habits of Schizothorax Niger In Dal-Lake, From Kashmir. Journal of Advanced Zoology, 45(3).

Qadri, S., Shah, T.H., Balkhi, M.H., Bhat, B.A., Bhat, F.A., Najar, A.M., Asmi, O.A., Farooq, I. and Alia, S., 2015. Absolute and Relative Fecundity of Snow Trout, Schizothorax curvifrons Heckel, 1838 in River Jhelum (Jammu &Kashmir). SKUAST journal of Research, 17(1), pp.54-57.

Qi, D. L., Chaȯ, Y., Guo, S. C., Zhao, L. Y., Li, T. P., Wei, F. L. & Zhao, X. Q. (2012). Convergent, parallel and correlated evolution of trophic morphologies in the subfamily Schizothoraci-nae from the Qinghai-Tibetan Plateau. PLoS One, 7, e34070.

Raina, H.S. and T. Petr, 1999. Coldwater fish and fisheries in the Indian Himalayas: lakes and reservoirs. p. 64-88. In T. Petr (ed.) Fish and fisheries at higher altitudes: Asia. FAO Fish. Tech. Pap. No. 385. FAO, Rome. 304 p.

Raina, H.S., 1977. Observations on the fecundity and spawning behaviour of Schizothorax esocinus Heckel from Dal Lake, Kashmir. Indian journal of fisheries, 24(1 & 2), pp.201-202.

Rampal, C. 1967. Studies on the life history of five Kashmir fishes together with fecundity and breeding habits. Ph.D. thesis, J&K University of Kashmir.

Riede, K., 2004. Global register of migratory species - from global to regional scales. Final Report of the R&D-Projekt 808 05 081. Federal Agency for Nature Conservation, Bonn, Germany. 329 p.

Sabha, K.K., Najar, A.M., Bhat, F.A., Shah, T.H., Balkhi, M.H. and Faisal, R., 2017. Reproductive biology of (Schizothorax niger) snow trout in Nigeen Lake Kashmir.

Sehgal, K.L., 1999. Coldwater fish and fisheries in the Indian Himalayas: rivers and streams. Fish and fisheries at higher altitudes: Asia. Food and Agriculture Organization of the

United Nations Technical Paper, 385, pp.41-63.

Shrestha, T.K., 1990. Resource ecology of the Himalayan waters. Curriculum Development Centre, Tribhuvan University, Kathmandu, Nepal. 645 p.

Silas, E.G., 1960. Fishes from the Kashmir valley. Journal of Bombay Natural History Society, 57(1), pp.66-77.

Sunder, S. 1984. Dietary habits of Schizothorax curvifrons (Heckel) from River Jehlum Kashmir. Himalayan Research and Development, 3(11): 17-23.

Sunder, S. and Bhagat, M. J. 1979. A note on the food of Schizothorax plagiostomus (McClelland) in the Chenab drainage of Jammu Province during 1973-74. J. Inland Fish. Soc. India, 11: 117-118.

Sunder, S. and Subla, B.A. 1985. Food of Juveniles of Schizothorax curvifrons (Heckel). Bull. Env. Sci., 2(1): 34-36

Sunder, S., Bhagat, M.J., Joshi, C.B. and Ramakrishna, K.V. 1979. On the fishery and some aspects of biology of Schizothorax niger from the Dal Lake. Journal of Inland Fisheries Society of India, 11:82-92.

Syed, T.M., 2024. Ecological Adaptations and Conservation Challenges of Schizothoracids in High Altitude Environments: A Mini Review.

Talwar, P. K. and Jhingran, A. G. 1991. Inland fishes of India and adjacent countries. Oxford Publication, New Delhi, India, 1158 p.

Walker, K.F. and H.Z. Yang, 1999. Fish and fisheries in western China. FAO Fish. Tech. Pap. 385:237-278.

Wang, M.M., Zhang, J.Y. and Zhao, Y.J. 2022. Morphological description and molecular identification of Myxobolus dajiangensis n. sp. (Myxozoa: Myxobolidae) from the gill of Cyprinus carpio in southwest China. PeerJ, 10: p.e13023.

Wani, I.F., Shah, T.H., Bhat, F.A., Bhat, B.A. and Balkhi, M.H., Assessment of gonadosomatic index and fecundity of Schizothorax niger in Dal Lake, Kashmir.

Wu, Y. and C. Wu, 1992. The fishes of the Qinghai-Xizang Plateau. Sichuan Publishing House of Science & Technology, Chengdu, China. 599 p.

Yonezawa, T., Hasegawa, M. and Zhong, Y., 2014. Polyphyletic origins of schizothoracine fish (Cyprinidae, Osteichthyes) and adaptive evolution in their mitochondrial genomes. Genes & genetic systems, 89(4), pp.187-191.

Yousuf, A. R. 1996. Fishery resource of Kashmir. In: A. H. Khan and A. K. Pandit (eds.) Ecology, Environment and Energy. University of Kashmir. pp. 75-120.

Yunfei, W. 1987. On the present status of Cyprinid fish studies in China. In: Seven, O. Kullander and B. O. Fernholm (eds.) Proc. 5th Cong. Eur. Icth. Swedish Museum of Natural History. pp. 43-48.

Zhang, C., Tong, C., Ludwig, A., Tang, Y., Liu, S., Zhang, R., Feng, C., Li, G., Peng, Z. and Zhao, K., 2018. Adaptive evolution of the Eda gene and scales loss in schizothoracine fishes in response to uplift of the Tibetan plateau. International Journal of Molecular Sciences, 19(10), p.2953.

6

Biology and Ecology of Salmonid Fishes in Kashmir

Arizo Jan*, Tasaduq Hussian Shah, Mir Iqra Farooq, Hudisa Banoo and Syed Talia Mushtaq

Division of Fisheries Resource Management, Faculty of Fisheries, SKUAST-Kashmir

Abstract

Salmonids include a diverse group of ray-finned, coldwater fish, comprising iconic species such as trout, salmon and char which are of economic, ecological and cultural significance. Salmonidae, which originated from the Order Salmoniformes, includes more than 200 species spread throughout the subarctic and temperate waters of the Northern Hemisphere. This chapter addresses the taxonomy, ecology, culture and conservation challenges of these representative species with particular emphasis on the trout introduction and propagation in Kashmir, India. Kashmir has flourished as trout culture hub due to its conducive climate and abundance of water resources, benefitting from advanced aquaculture techniques and improved infrastructure supported by international collaborations. Despite this, salmonid conservation status, including rainbow trout, continues to be precarious as habitat degradation, pollution, and climate change worsen. This chapter has highlighted the intricate relationship between human activities and the ecological health of freshwater ecosystems, emphasizing the critical importance of securing the future of rainbow trout and other salmonid species around the world by ensuring sustainable management practices.

Keywords: *Salmonids, Salmoniformes, Coldwater fish, Rainbow trout, Taxonomy, Conservation*

Introduction to the Salmonid Family (Trout's and Salmon)

The Salmonidae constitute the only extant family of ray-finned fish in the order Salmoniformes, with 11 genera and more than 200 species collectively referred to as "salmonids" or "salmonioids". This family includes all coldwater mid-level predatory fish found in subarctic and cool temperate waters in the Northern Hemisphere, such as char, graylings, freshwater whitefish, taimens, lenoks, and salmon (both Atlantic and Pacific species). In addition to being the eponym of its genus, the Atlantic salmon (*Salmo salar*), has also been given the names of its family and order as well. Among teleost fish, salmonids have a relatively primitive appearance, with pelvic fins in the rear and an adipose fin at the rear. There is a single row of sharp teeth in their mouth, along with rounded scales and forked tail fins (Mc Dowell, 1998). Salmonids vary greatly in size, and the smallest reach only 13 cm (5.1 in) when adults, while the largest reaches 2 m (6 ft 7 in) (Pauly *et al.*, 2008). Salmonids are migratory fish that spawn in the shallow gravel beds of freshwater headstreams, develop into juveniles in rivers, creeks, small lakes, and wetlands, and then migrate downstream to spend the majority of their adult lives in larger bodies of water upon reaching maturity. Often euryhaline, salmonid species migrate to the sea or brackish estuaries as soon as they reach adulthood. Just for breeding purposes, they migrate back to the upper streams. Some freshwater salmonids that migrate only between lakes and rivers are referred to as potamodromous, whereas the life cycle of these sea-run salmonids is known as anadromous. In the middle food chain, salmonids are carnivorous predators that eat smaller fish, crustaceans, aquatic insects and larvae, tadpoles, and occasionally fish eggs (even those belonging to their own species), which are then eaten by larger predators. Salmonids are regarded as keystone organisms that contribute biomass to freshwater and terrestrial ecosystems as a result of their mass migration from oceanic to inland waterbodies. Three lineages of salmonids are currently recognized, each classified as a subfamily: Coregoninae (freshwater whitefish), Thymallinae (graylings), and Salmoninae (trout, salmon, char, taimens, and lenoks). Each of the three lineages is generally accepted to belong to a monophyletic group based on derived traits (McPhail and Strouder, 1997).

Salmons and trouts are well-known for their economic importance as both sport fish and commercial products. Trout and salmon stocks are stocked worldwide each year for sport and commerce. Millions of salmon and trout are hatched, raised, and stocked every year as part of government investment programs. To meet market and restaurant demands for trout, a large private industry has developed—especially in Denmark, Japan, and the U.S. Scientists are confronted with the challenge of perpetuating and increasing the abundance of salmon and trout because of the increasing human population and industrial and agricultural demands on rivers. Among heavily populated regions, the demand for trout outstrips the supply. As a result, state and federal agencies in the United States have initiated massive programs to raise and stock trout in heavily fished waters. The artificial abundance of trout, however, is not a good substitute for the natural abundance of the fish.

History of the introduction of Salmonid fish in the Kashmir Valley

Several reasons have been used to introduce exotic fish species to many parts of the world, including: to improve local fishery potential, to increase species diversity in aquatic systems, to fish for sport, to keep aquariums, and to control undesirable organisms. The introduction of brown trout and rainbow trout was primarily aimed at satisfying sport fishing demands. Streams of Kashmir are now home to numerous self-sustaining brown trout populations. A few rainbow trout have been found in streams, but they are confined to fish farms (Petr, 1999). Furthermore, brown trout prey on the Schizothoracines' younger stages, causing a sharp decrease in their populations (Sehgal, 1999). It has been common practice among nations to culture Salmonids for many years, making them one of the most cultured fish species across the globe (Lee and Donaldson, 2001). From Europe, North America and Canada, five Salmonid species were introduced into the Himalayan range from 1905 to 1969. These included brown trout, rainbow trout, eastern brook trout (*Salvelinus fontinalis*), splake (a hybrid trout between lake and brook trout), as well as a land-locked Atlantic salmon (*Salmo salar*). The brown trout population in streams of the Himalayas is one of the best-established of these species. There are five major trout-producing states in India, each with a well-developed infrastructure and access to water in the required quantity and quality including Himachal Pradesh, Jammu and Kashmir, Tamil Nadu, Uttarakhand and Kerala. Over the past few decades, rainbow trout have been systematically relocated throughout the world. They are now well established in countries such as South America, Japan, China, Europe, Africa, Australia, New Zealand, India and Pakistan.

Introducing trout to Kashmir began in 1898 when F.J. Mitchell, a Scotsman, Pandit Sodhama Miskeen, and Khwaja Gafarjoo collaborated to introduce the fish. First batch of 10,000 eggs arrived from the United Kingdom in 1899 as a gift from the Duke of Bedford, who provided the trout seed. A first attempt at introducing trout to Kashmir failed. But in December 1900, a second shipment of 1800 trout fry arrived from Scotland in excellent condition. The transport of rainbow trout to Kashmir is credited to F. J. Mitchell, the "father of Kashmir trout fishing." 1000 of them were relocated to Panzagam Dachigam (Harwan), about 24 km from Srinagar while the other 800 were raised in his premises at Baghi DilawarKhan, close to Khanqah-e-Moulla. Known as "Gada Sodhama" (Sodhama, the Fisherman), Mr. Sodhama built a hatchery in Achabal, Anantnag District, from which brown trout eyed-ova were distributed across the Kashmir (Kaul, 2015). This led to new trout beats being developed in the valley's snow-fed streams like Bringhi, Lidder, Sindh, and Madhumati. Verinag, Kokernag, Achhabal, Gangabal, Vishensar, Kisenar, Satsar, Gadsar, Sheshnag, and Kounsarnag were among the high-altitude lakes stocked with trout.

Fish consumption has helped people overcome animal protein deficiency all winter long and even during dry seasons. Currently, Jammu and Kashmir produces 21.35 thousand tonnes of fish per year (Jammu and Kashmir Fisheries Department, 2019-2020) and imports fish to satisfy its consumption demand. With the establishment of Mother Trout Fish Farming Project at Kokernag under the assistance of the European Union, the fishing of trout in Jammu & Kashmir has gained a substantial boost in the last thirty years. Besides the existing Trout Hatchery, the Department has built a new one at Harwan, Srinagar to produce Brown trout and Rainbow trout seeds. The new trout rearing units are also being established. The culture of trout uses modern innovation to breed and rear the fish to create a higher endurance level at different stages. A very well-equipped Trout feed mill imported from Holland has been set up at Manasbal so that the Department can produce quality fish feed that will result in improved conversion ratios and a healthy merchantable stock (Jammu and Kashmir Fisheries Department, 2020).

Taxonomy of Rainbow Trout and Other Salmonids

Various forms of rainbow trout are named *Oncorhynchus mykiss*. There are three different ecological forms of rainbow trout:

1) Sea-run or anadromous, 2) Stream resident, and 3) Lake and reservoir dwelling.

Among its anadromous forms, steelhead are classified as *Oncorhynchus mykiss irideus*. Many subspecies of rainbow trout have been described based on their appearances and life histories, including coastal rainbow trout, redband trout, and natives of the Gulf of California (Behnke, 2002). A stream-resident rainbow trout's back and sides are typically silvery-gray to greenish brown with a white belly. There is a pink or reddish lateral stripe along the entire length of the rainbow trout's body, sometimes accompanied by lavender or orange overtones. A rainbow trout has various spotting patterns and body colorations. In general, the caudal fins have rows of dark spots, and the dorsal and adipose fins have spots. The lower fins are pale-pink and spotless. The head and sides tend to be marked with small blackish spots. The side stripes of the males turn intensely red at spawning time.

Habitat and Feeding Behavior

There are several factors that determine the natural diet of rainbow trout, including their size, age, and habitat. Hoitsy (2002) describes rainbow trout as being a voracious and aggressive feeder. The fish eat almost anything that falls into the water, just like opportunistic feeders. As with insects, these include adult beetles (Coleoptera), flies (Diptera), ants (Formicidae) and moths and butterflies (Lepidoptera) larval or egg stages (Montgomery and Bernstein, 2008). Despite being predominantly carnivorous, rainbow trout have few anatomical attributes for acquiring and digesting their food. They also lack elaborate structures for capturing, holding, and swallowing prey, aside from their simple and small teeth. Salmonids have Y-shaped stomachs and a wide oesophagus through which they swallow their food whole. Salmonid taxonomic significance often rests upon the abundance of pyloric caeca in the midgut at the pyloric end. It is difficult to see the pancreas through the pyloric caeca due to its distribution in the fat and connective tissue. On larger specimens, the bile duct extends from the middle lobe of the liver to the upper midgut. Midgut and hindgut do not have a distinct separation, the kidney occupies most of the visceral cavity just dorsal to the swim bladder. Overall, rainbow trout's are usually representative of salmonids. Gut length (esophagus to anus) in general ranges from 0.6 to 0.8 times body length, which is barely longer than teleost's.

The growth of rainbow trout is mainly affected by four essential habitat factors. These include the natural food abundance and the basic qualities of the water. Cold water: Rainbow trout is a typical cold water fish, pristine water: Sharp vision is essential for trout feeding success. Dissolved oxygen: To maintain efficient respiration, water should maintain high concentrations of DO. Pure water: The water should be devoid of any toxic solid and gaseous waste products that generate during the processes of metabolism and respiration. According to Raleigh *et al.* (1984), rainbow trout prefer water that is between 12 and 18°C, with temperatures over 25°C potentially leading to fish death. Given that their growth is negligible during prolonged cold water temperatures, researchers have discovered that adult rainbow trout reduce their feeding during the winter to maintain a minimum level of metabolic activity (Raleigh *et al.*, 1984). Apart from alterations in metabolism, the density and depth of the pool are crucial factors that affect the survival of overwintering trout. The majority of the fish habitat in most small streams consists of pools, and the depth of the pools appears to have a significant impact on the variety and abundance of trout (USFS 1994). Quality pools should generally have a depth of 1-2 meters (Raleigh *et al.*, 1984; USFS, 1994). The Kashmiri rainbow trout spawns from November and lasts till February. Incubation periods for rainbow trout range from 28-49 days and vary inversely with water temperature.

Culture Practices

Sport fishing, meat quality, and simplicity of culture of propagation make rainbow trout culture more popular than other Salmonid species. Presently, gametes derived from fish brood stocks are increasingly being recognized as being important in ensuring the viability of hatchlings in culture practices (Kjorsvik *et al.*, 1990; Billard *et al.*, 1995; Yaron, 1995). Fish age has been found to have an impact on hatchling survival rates and gamete quality. In contrast to younger, smaller females, older, larger rainbow trout lay larger eggs (Gall, 1974). There is a long history of Salmonid culture in Europe and North America and it began a long time after carp culture (Pillay, 1990). The brown trout is the first European Salmonid fish to be reared and reproduced artificially, but these days, rainbow trout farming is receiving much attention. The rainbow trout is a hardy fish that can tolerate a wide range of environmental conditions and accept artificial feed well. Therefore, it is ideal for fish culture. A trout is eurythermal, which allows it to withstand higher temperatures, have a shorter incubation period, grow faster and are more resistant to certain diseases. Temperature and food availability affect growth and maturity, thus causing a variation in maturity age.

Here are a Few Factors that Contributed to Trout Culture's Early Success

- The species can adapt well to a wide range of environmental conditions,
- Both sexes are capable of maturing easily in captivity,
- Both sexes' gametes could easily be stripped,
- External fertilization can be performed using any container,
- Upon hatching, the larva is well-developed and readily accepts artificial feed.

A trout's breeding season begins in November. As trout do not naturally spawn in culture frameworks, the young are artificially spawned in incubation facilities or gathered from wild populations. Since the climate is favorable in J&K, trout production is extremely high, with a low mortality rate. There has been an increase in the production of trout (biomass) from 482 tonnes in 2017 to 650 tonnes in 2020, as well as an increase in the production of trout seed from 132 lacs in 2017 to 147.50 lacs. There are a number of environmental factors, topography, and climate factors that make Kashmir a suitable place for breeding, rearing, and producing trout. In the Kashmir valley, glacier-fed streams, mountains, lakes, springs, and other areas promote sound trout breeding, while the physico-chemical parameters of the cold water are very conducive to trout production. The endurance pace of trout eggs in Kashmir is about 10% more than the western average of 46% (Hassan and Pandey, 2012). Five seed producing incubation centers and breeding projects are located in Jammu and Kashmir: Laribal (Srinagar), Kokernag (Anantnag), Shokababa (Baramulla), Mammar (Ganderbal) and Phalni (Rajouri). Breeding is conducted in controlled conditions by stripping eggs, from which fingerlings are raised in designated hatcheries, and then supplied to large and small trout rearing units located all over the state. Breeding season for Kashmiri rainbow trout runs from November to February. Rainbow trout can incubate for up to 49 days, depending on the temperature of the water.

Production Conditions

Rainbow trout production should be conducted under optimal or near-optimal conditions for all age groups. As a result, it is important to understand the fish's fundamental requirements and the perfect farming conditions.

1. **Water pH:** Depending on the stage of development of the fish, rainbow trout have varying tolerances to unfavorable pH* conditions. Rearing water's ideal and acceptable pH ranges vary as well. The ideal pH range for developing embryos and fry is limited to 6.5 to 8, but the acceptable pH range is likewise constrained. For older fish, there are wider pH ranges that are both optimal and acceptable.
2. **Temperature of water**: The ideal, acceptable, and lethal ranges for water temperature differ as well based on the fish's developmental stage. Water temperatures between 7 and 18 °C are ideal for rainbow trout feeding. Above and below this range, they lose appetite. At too low or too high a water temperature, fish stop feeding and rainbow trout's appetite and feed intake begin to intensify as the water temperature increases. However, this behavior only lasts until around 18 °C. The fish abruptly decrease their appetite and feed intake once they reach this temperature and cease feeding. Remember that there is an inverse relationship between feeding intensity and utilization to gain an understanding of how feeding intensity affects utilization of consumed feed. Although the digestion of feed is less complete at 18 °C, rainbow trout are still willing to feed very intensively at this temperature. Depending on the species, the water temperature where trout grow the best is usually between 13 °C and 15 °C (Baldwin, 1957) or 15 °C (Molony, 2001). Hence, rainbow trout use their feed optimally and have a maximum appetite within this range of temperatures.
3. **Dissolved oxygen:** Content of Oxygen (O_2) in water ensures the respiration of aquatic animals and plants. The standard unit of measurement for DO is milligrams of oxygen per liter of water (mg/l). The maximum oxygen content of water depends on its temperature, as water can dissolve only a certain amount of oxygen, which is determined by the atmospheric partial pressure of oxygen. In water, temperature and DO content are inversely correlated. A lower DO content is the outcome of a higher

temperature, and vice versa. Water is 100% saturated with oxygen at its maximum oxygen content, and any extra oxygen promptly escapes into atmosphere. Oxygen concentrations in water are optimal and acceptable depending on the fish's development stage. For example, rearing water should have an oxygen content near to saturation (100%). Water with a low oxygen content of 5 to 6 mg/litre is suitable for incubating eggs and developing fry; for older age groups, the acceptable low oxygen content may be as low as 4-5 mg/litre. Furthermore, fish take in more oxygen both during and after feeding. During these times, there will be a temporary rise in the need for oxygen.

4. **Water supply:** Maintaining an oxygen-rich, clean, fresh water supply is essential for replenishing used water in rearing devices. The quantity of water provided varies depending upon the age and number of developing fish. Water's oxygen content has an impact on the number of eggs, fry, and growing fish per unit area of the rearing devices. Colder water slows metabolism and, therefore, respiration, while warmer water intensifies it. As a result, the quantity of water needed in each water temperature to raise the same number of embryos, fry, and fish will vary. Depending on the water temperature, less water may be supplied at low temperatures, but more water may be supplied at higher temperatures. It is measured as the flow rate, which is the quantity of water needed for 10,000 or 1,000 specimens of eggs, fry or fish. Either liters per minute (liters/min) or liters per second (liter/s) are used to express flow rates. An alternative method of characterizing the amount of water supplied is an exchange rate. It is expressed in terms of how much water is exchanged per hour or day. Concrete or lined tanks can have a higher water supply than earth ponds, so fish can be more dense in these devices.

Reproductive Ecology and Early Life Stages

A trout or salmon (Salmonidae) can migrate to the sea and then return to freshwater to spawn, or they can spend their entire life in freshwater. Species classified as anadromous are those that spend the majority of their lives in the sea but move to freshwater habitats to reproduce. On the other hand, freshwater-dwelling species that migrate to the marine environment for reproduction are known as catadromous species. Depending on the species, gravelly shoals in small streams are the typical location for salmon and trout spawning, which occurs in the spring or fall. The physical characteristics that develop during spawning are the easiest way to identify male and female Salmonids. Males who are older and larger tend to develop kypes, which is characterized by a hooked lower jaw. In contrast, the snouts of female trout tend to be rounded.

In addition, males usually have a more laterally compressed body shape than females, who tend to have rounder bodies. Males may also have brighter coloration than females during the spawning season. In order to create their nests in the gravel, females lay on their sides against the bottom, propel themselves forward with great force, and flush out stones with their bodies and fins. A redd is the name for a female's nest. The eggs land in these gaps created by the loosening of the gravel inside the redd, which lets cold, pure, oxygenated water pass through. Before she departs, the female may add more gravel to the redds to shield the eggs from the sun and potential predators. Depending on the water temperature, eggs can hatch in 4 to 10 weeks. Until the yolk sac is absorbed, young trout remain in gravel. They then transition into running streams. Trout populations that are reproducing are believed to be an indicator of clean, oxygenated, unpolluted water. Figures 1 and 2, which represent the developmental stages of brown trout and rainbow trout, respectively, are displayed below.

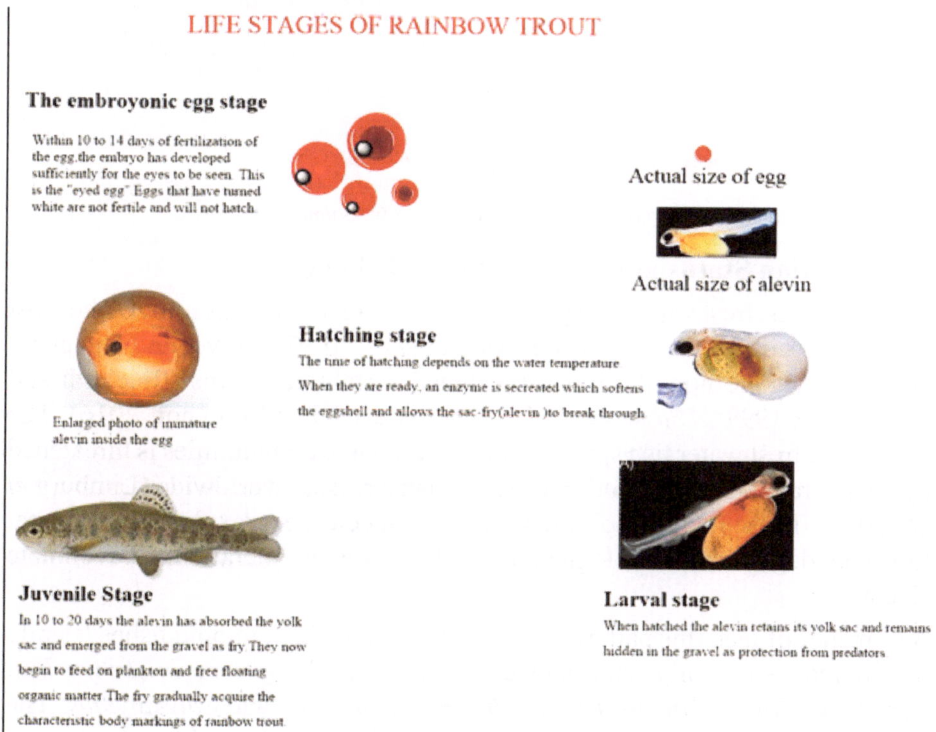

Fig. 1: Illustration of developmental stages in rainbow trout, *Oncorhynchus mykiss*.

Fig. 2: Illustration of developmental stages in brown trout, *Salmo trutta fario*.

Conservation Status and Management Challenges

In recent times, freshwater ecosystems have been recognized as some of the most threatened ecosystems worldwide (Richter *et al.*, 1997; Strayer and Dudgeon, 2010). Fishes are now the most threatened vertebrate groups (Ricciardi and Rasmussen, 1999; Vörösmarty *et al.* 2010; Darwall and Freyhof, 2016). The survival of freshwater fish species and entire aquatic communities is threatened by a wide range of direct and indirect human stressors worldwide (Limburg *et al.*, 2011). These include pollution, invasive species, overexploitation, habitat loss and degradation, biological flow diversion or alteration, and climate change.

As a matter of fact, this pattern is especially evident in Salmonid fishes (family Salmonidae, subfamily Salmoninae, belonging to the genera *Oncorhynchus, Salvelinus, Salmo, Hucho, Parahucho, Brachymystax,* and *Salvethymus*. The salmonid family is a cold-water species that lives exclusively in freshwater ecosystems (typically trout and char), though there are also species that migrate from the ocean to freshwater to spawn (e.g., Atlantic salmon salar, Pacific salmon *Oncorhynchus* spp.). Aside from providing recreation and food to millions of people, these cold-water specialists play an important role in ecosystem health and functioning (Holmlund and Hammer, 1999). As a result of the fact that they need cold, clean, complex, and connected habitats to survive and persist, they are extremely sensitive to human disturbances (Gresswell *et al.*, 1997; Haak and Williams, 2013; Hauer *et al.*, 2016), and all of these attributes have been substantially altered and degraded by humans. Despite their widespread significance as symbols of society and biodiversity, many trout and char species and lineages are endangered worldwide, and some urgently need conservation efforts in order to reverse their rapid extinction (Muhlfeld *et al.*, 2018). At this point, an extensive understanding of their global status, threats, and conservation concerns is lacking. There have been alarming declines in native trout and char diversity throughout the world as a result of several human impacts on freshwater habitats and populations. Among these impacts are invasive species (69% of species), overfishing (47%), pollution (33%), and dams (28%) mentioned in the IUCN assessments. There are other threats, such as deforestation, agriculture, grazing, water abstraction, roads, and extractive industries (e.g., mining), that affect 6 to 22% of trout and char species. In particular, char and trout are highly sensitive to anthropogenic stressors and climate change, which often increase the likelihood that populations will decline and extinctions will occur (Kovach *et al.*, 2017). As environmental change accelerates, we must protect ecological and genetic diversity to ensure long-term viability and resilience of these fishes. There are a variety of innovative conservation approaches including re-connecting rivers with floodplains, creating native fish refuges, restoring habitat diversity, and reducing invasive species, including stocking nonnative trout. Furthermore, comprehensive, coordinated, and comparable approaches are urgently needed for assessing conservation status and delineating conservation units worldwide, particularly for species with scarce data. This conservation goal cannot be achieved until we identify and then address the root causes of threats to ensure that these valuable fishes remain on the landscape into the 22nd century and beyond.

Conclusion and Way Forward

Salmonids-which include well-known species like salmon and trout-are essential to freshwater ecosystems all over the world. They are keystone species because of their importance in ecological, economic, and cultural domains. But in the face of growing environmental changes and human impacts, these fish face many conservation challenges. Because of their anadromous and potamodromous life cycles, Salmonids are especially vulnerable to pollution, invasive species, overexploitation, habitat degradation, and climate change.

Together, these elements pose a threat to their biodiversity and survival. In order to protect Salmonid populations, urgent and coordinated action is required as freshwater ecosystems around the world come under increasing strain. The introduction of European brown and rainbow trout to Kashmir has improved sport fishing and improved local economies and food security. These introductions have been successful, but there have been ecological drawbacks as well, like competition with native species and changes to aquatic habitats. Maintaining local biodiversity while managing these introductions sustainably is still a major challenge. Integrated conservation strategies that give habitat restoration, reducing human impacts, and sustainable management practices top priority are essential for the survival of Salmonids.

References

Baldwin, N.S. 1957. Food consumption and growth of brook trout at different temperatures. Transactions of the American Fisheries Society, 86: 323–328.

Behnke, R. J. 2002. Trout and salmon of North America. First edition. The Free Press, Simon and Schuster Inc., New York pp. 67-135.

Billard, R., Cossin, J., Crim, L.W. and Suquet, M. 1995. Sperm physiology and quality. In: Bromage, N., Roberts, R. (Eds.), Broodstock Management and Egg and Larval Quality. Blackwell, Oxford, pp. 25-52

Darwall, W. R. T., and J. Freyhof. 2016. Lost fishes, who is counting? The extent of the threat to freshwater fish biodiversity. Pages 1–36 in G. P. Closs, M. Krkosek, and J. D. Olden, editors. Conservation of freshwater fishes. Cambridge University Press, Cambridge, UK.

Froese, Rainer, and Daniel Pauly, eds. (2008). "Salmonidae" in FishBase. December 2008 version.

Gall, G.A.E. 1974. Influence of size of eggs and age of female on hatchability and growth of rainbow trout. California Fish and Game 60(1):26-35.

Gresswell, R. E., J. Kershner, J. B. Dunham, and R. E. Gresswell. 1997. Ecology and management of potamodromous salmonids: introduction to ecology and management of potamodromous salmonids. North American Journal of Fisheries Management 17:1027–1028.

Haak, A. L., and J. E. Williams. 2013. Using native trout restoration to jumpstart freshwater conservation planning in the Interior West. Journal of Conservation Planning 9:38–52.

Hassan, N. and Pandey, D. N. 2012. Present status of trout fisheries in Jammu and Kashmir. IOSR Journal of Pharmacy 2(5): 35-37.

Hauer, F. R., H. Locke, V. J. Dreitz, M. Hebblewhite, W. H. Lowe, C. C. Muhlfeld, C. R. Nelson, M. F. Proctor, and S. B. Rood. 2016. Gravel-bed river floodplains are the ecological nexus of glaciated mountain landscapes. Science Advances [online serial] 2:e1600026.

Hoitsy, G. 2002. A Pisztráng tenyésztése és horgászata. 152 oldal

Holmlund, C. M., and M. Hammer. 1999. Ecosystem services generated by fish populations. Ecological Economics 29:253–268.

Kaul, B.L. 2015. Trout fish culture has a history in Kashmir. Greater Kashmir. https://www.greaterkashmir.com/more/trout-fish-culture-has-a-history-in-kashmir (Accessed on 12 July 2023)

Kjorsvik, E., A. Mangor-Jensen and I. Holmefjord. 1990. Egg quality in fishes. Advances in Marine Biology26: 73-113.

Kovach, R. P., R. Al-Chokhachy, D. C. Whited, D. A. Schmetterling, A. M. Dux, and C. C. Muhlfeld. 2017. Climate, invasive species and land use drive population dynamics of a cold-water specialist. Journal of Applied Ecology 54:638–647.

Lee, C.S. and Donaldson, E.M. 2001. General discussion on "Reproductive biotechnology in finfish aquaculture.Aquaculture197: 303-320.

Limburg, K. E., R. M. Hughes, D. C. Jackson, and B. Czech. 2011. Human population increase, economic growth, and fish conservation: collision course or savvy stewardship? Fisheries 36:27–35.

McDowell, Robert M. 1998. Paxton, J.R.; Eschmeyer, W.N. (eds.). Encyclopedia of Fishes. San Diego: Academic Press. pp. 114–116. ISBN 978-0-12-547665-2.

McPhail, J.D., Strouder, D.J. 1997. Pacific Salmon and Their Ecosystems: Status and Future Options. The Origin and Speciation of *Oncorhynchus*. New York, New York: Chapman & Hall.

Molony, B. 2001. Environmental requirements and tolerances of rainbow trout (*Oncorhynchus mykiss*) and brown trout (*Salmo trutta*) with special reference to Western Australia: a review. Fisheries Research Report No. 130. Perth, Australia, Fisheries Research Division.

Montgomery, W.L. & Bernstein, Y. 2008. Rainbow trout (*Oncorhynchus mykiss*): a technical conservation assessment. Rocky Mountain Region, Species Conservation Project. USDA Forest Service.

Muhlfeld, C. C., D. C. Dauwalter, R. P. Kovach, J. L. Kershner, J. E. Williams, and J. Epifanio. 2018. Trout in hot water: a call for global action. Science 360:866–867.

Petr, T.1999. Fish and fisheries at higher altitudes: Asia. FAO Fisheries

Pillay, T.V.R. 1990. Aquaculture: principles and practices. Oxford Fishing News Books [A well-illustrated book on Aquaculture] pp. 576.

Raleigh, R. F., Hickman, T., Solomon, R. C. and Nelson, P. C. 1984. Habitat suitability information: Rainbow trout. US Fish and Wildlife Service FWS/OBS 82(10): 60-64.

Ricciardi, A., and J. B. Rasmussen. 1999. Extinction rates of North American freshwater fauna. Conservation Biology 13:1220–1222.

Richter, B. D., D. P. Braun, M. A. Mendelson, and L. L. Master. 1997. Threats to imperiled freshwater fauna. Conservation Biology 11:1081–1093.

Sehgal, K. L. 1999. Coldwater fish and fisheries in the Indian Himalayas: Rivers and streams. FAO: ICAR.

Strayer, D. L., and D. Dudgeon. 2010. Freshwater biodiversity conservation: recent progress and future challenges. Journal of the North American Benthological Society 29:344–358

technical paper, No.385. Rome, FAO.1999.pp. 304.

USFS. 1994. Section 7 fish habitat monitoring protocol for the upper Columbia River Basin. Pacific Northwest Region, Intermountain Region, Northern Region.

Vörösmarty, C. J., P. B. McIntyre, M. O. Gessner, D. Dudgeon, A. Prusevich, P. Green, S. Glidden, S. E. Bunn, C. A. Sullivan, C. R. Liermann, and P. M. Davies. 2010. Global threats to human water security and river biodiversity. Nature (London) 467:555–561.

Yaron, Z. 1995. Endocrine control of gametogenesis and spawning induction in the carp. Aquaculture, 129: 49-73.

7

Study on the Biological Characteristics Reproductive Behaviour and Habitat Adaptation of Chinese Carp Species in Kashmir's Aquatic Ecosystems

Bisma Khan**, *Tasaduq H. Shah, Rinkesh Nemichand Wanjari and Zaheer Abass

Division of Fisheries Resource Management, Faculty of Fisheries, SKUAST-Kashmir Rangil, Ganderbal, Jammu and Kashmir

Abstract

The global distribution and extensive introduction of the Chinese carp (Cyprinus carpio) underscore its status as one of the most widely introduced fish species worldwide. Originally introduced primarily for aquaculture purposes, the species has established populations in over 100 countries, often leading to significant environmental impacts upon invading freshwater ecosystems. Its adaptability to varying environmental conditions, including temperature and turbidity, coupled with prolific breeding capabilities, facilitated its rapid establishment in diverse inland waters such as rivers, lakes, streams, canals, and wetlands. In India, the species has become highly abundant, serving both as a crucial food fish and a target for recreational fishing in regions like Kashmir. Alongside the Chinese carp, several other introduced species, including various cyprinids (such as Carassius carassius and Ctenopharyngodon idella), salmonids (Salvelinus fontinalis, Onchorhyncus nerka, and Salmo salar), and others like Tilapia mossambica and Gambusia affinis, have also been documented post-independence. These introductions, while often beneficial economically, highlight the complex ecological implications associated with deliberate species introductions in non-native habitats.

Keywords: *Chinese carp, Aquaculture, Introduced species, Environmental impacts, Non-native.*

Introduction

Any species that has been accidentally or purposely brought outside of its current range by humans is considered an introduced (or exotic) species (Kottelat and Whitten, 1996). According to the International Union for Conservation of Nature and Natural Resources (IUCN), an alien species is considered invasive when it establishes itself in natural or semi-natural ecosystems or habitat, causes change, and endangers the biological diversity of the host species. These invasives encompass all classes of living things and are widely dispersed throughout the world's ecosystems (Raghubanshi *et al.*, 2005). Numerous reasons led to the introduction of exotic fish species in various parts of the world: (a) increasing the potential for local fisheries and diversifying species in aquatic systems; (b) promoting sport fishing; (c) maintaining aquariums; and (d) managing undesired organisms like mosquitoes, etc. (Biju Kumar, 2000). Furthermore, floods and a variety of other factors may allow exotic fish imported into one nation to spread to its neighbors. According to Welcome (1988), 168 fish species from 37 families have been introduced worldwide outside of their native area. At least 67 of these species have established themselves in various bodies of water, with 27 of them turning out to be serious pests. Approximately 20 percent of freshwater fish species worldwide are either extinct or in danger of going extinct (Moyle and Leidy, 1992). Over time, the introduction of new species might prove to be a very harmful issue. Nyman (1991) stated that the introduction of alien species may cause the aquatic environment to undergo permanent alterations and perhaps cause the extinction of some species.

As a result, the careless introduction and movement of aquatic species, especially fish, caused much worry across the world since it led to several issues, including the extinction of native species. Fish native to the area compete with exotics for food and habitat. They might kill local fish, bring in new illnesses and parasites, create hybrids, pollute the genetic pool, and lower the standard of the aquatic ecosystems. The loss of priceless biodiversity will result from all of these (Nyman, 1991). Potential hazards include those that might affect not just biodiversity quality or quantity but also the socioeconomic facets of the human society that depend on aquatic environments for survival (Philipp *et al.*, 1995). Introductions are a legitimate way to increase aquaculture and fisheries productivity and economic advantage. Indeed, the primary driving force behind the migration of fish and shellfish across continents and even geographic regions has been aquaculture. To address the growing need for food and nutritional security, exotic species have been used to diversify fisheries and aquaculture. Species introductions and transfers from one hydrographic regime to another have defined the global aquaculture industry's growth. For instance, our well-known "composite fish culture" includes three exotic carps, namely common carp (*Cyprinus carpio*), grass carp (*Ctenopharyngodon idella*), and silver carp (*Hypophthalmichthys molitrix*), in addition to three major Indian carps, namely rohu (*Labeo rohita*), catla (*Catla catla*), and mrigal (*Cirrhinus mrigala*) (Raman *et al.*, 2013). In addition to these native carps, several more carps and suitable species were imported to fill the available niche and increase productivity. An alien species whose introduction does or is likely to cause economic or environmental harm or harm to human health" is the definition of "non-indigenous species" or "invasive species." The operation of ecosystems, their economic worth, and human health can all be impacted by invasive species, which can also change the ecological connections among native species. When a species is brought by humans to a place, region, or area where it was not previously found naturally (i.e., when it is not native), it is considered invasive if it can establish a br eeding population there on its own without additional human intervention and spreads widely throughout the new area (Singh *et al.*, 2010).

More than 3000 exotic fish species have been brought to India over the past few decades for aquaculture, sport fishing, ornamental fish keeping, mosquito control, and other purposes. In addition, a species of mollusk known as *Mytilopsis sallei*, which is native to Central America, was likely introduced to Indian waters through ship fouling. The natural water bodies of India are now home to several well-established exotic fish species. Reports from all over the world unequivocally demonstrate the extremely detrimental effects of exotic fishes on the aquatic ecosystems of India, even though not much research has been conducted on the effects of alien species in Indian waterways (Swain *et al.*, 2017). The following is a list of some significant fish introductions and their effects:

- The common carp is an example of a nuisance fish that has had a significant negative impact on several aquatic ecosystems.
- The introduction of silver carp to the Govindsagar Reservoir in Himachal Pradesh has decimated the once-thriving native Indian major carp population in the reservoir, which is now dominated by invasive silver carp. It creates muck in the water and obliterates the food and cover that other fish require by feeding on submerged vegetation and uprooting plants that support and nourish other aquatic animals. Additionally, the native populations of *Osteobrama belangeri* and *Schizothorax* sp. have been reported to be significantly impacted by the introduction of

Cyprinus carpio var. *specularis* into Dal Lake (J&K) and Loktak Lake (Manipur), respectively (Swain *et al.*, 2017).

- Another major worry is the introduction of the carnivorous and voracious Chinese bighead carp (*Aristichthys nobilis*) into the eastern and northeastern regions of India. This invasive fish poses a major danger to smaller native fish species and invertebrates if it establishes itself in natural water bodies (Joshi *et al.*, 2021).

Characteristics and Dynamics of Chinese carps

Grass carp (*Ctenopharyngodon idella*)

Habitat preferences and distribution

The Grass Carp (*Ctenopharyngodon idella*) is a Large Herbivorous Freshwater Fish Native to the Pacific Far East.

1. **Habitat:** Grass carp inhabit lakes, ponds, pools, and backwaters of large rivers. They prefer large, slow-flowing or standing water bodies with abundant vegetation. In the wild, they spawn in fast-moving rivers, and their eggs develop while drifting downstream, kept in suspension by turbulence.
2. **Distribution:** Their native range stretches from northern Vietnam to the Amur River on the Sino-Russian border. Grass carp have been introduced to various countries worldwide, including China, Taiwan, Russia, Singapore, Peru, United States, Mexico, Japan, United Kingdom, Netherlands, and more. They are often used for aquatic weed control due to their ability to consume vegetation.

Morphological Characters

The grass carp (*Ctenopharyngodon idella*) possesses several distinctive morphological features:

It has an elongated, cylindrical body with a slightly lengthened upper jaw and a short snout. The body color varies from light yellow to grey-green on the back and grey on the abdomen. The dorsal and pectoral fins are dark green, while other fins are light grey. Grass carp have large scales arranged along a straight lateral line (usually 37-42 scales). Their pharyngeal teeth are broad and ridged, arranged in a "2, 4-4, 2" formula. Adult grass carp can reach lengths of 3.5 to 4.8 times their body height. They typically mature at 3 to 6 years of age, with males maturing earlier than females. At maturity, their body weight ranges from 3 to 8 kg (Zhao *et al.*, 2020).

Feeding Adaptations and Dietary Habits

Grass carp (*Ctenopharyngodon idella*) are herbivorous freshwater fish with interesting feeding habits: They primarily feed on aquatic plants, including both higher aquatic plants and submerged terrestrial vegetation. Additionally, they may consume detritus, insects, and other invertebrates. Grass carp can eat up to three times their own body weight daily, making them voracious herbivores (Li *et al.*, 2023).

Reproduction and Early Life Stages

Grass carp spawn in turbulent rivers. Their eggs, which are slightly heavier than water, develop while drifting downstream, kept in suspension by turbulence. The eggs require long rivers for survival, and they are thought to die if they sink to the riverbed. After hatching, grass carp larvae continue to drift in the river. They have a short period of development during this drift phase. Eventually, they must swim from the river and find appropriate low- or no flow nursery areas. Gonadal maturation occurs in still waters. Spawning is associated with rising water levels and occurs in running water, where the eggs become pelagic. Grass carp mature at different ages depending on their location: Males typically mature one year earlier than females (Mohd *et al.*, 2023).

Growth Dynamics and Lifespan

Grass carp (*Ctenopharyngodon idella*) exhibit interesting growth dynamics and have a notable life span: On average, grass carp can live up to 14 years, but there have been reports of individuals living beyond 20 years under ideal conditions. These fish are known for their rapid development. Young grass carp stocked in the spring at around 20 cm (7.9 in) can reach over 45 cm (18 in) by fall. Their typical length ranges from about 60 to 100 cm (23.5 to 39.5 in), with a maximum length of 2.0 m (6.6 ft) and a weight of up to 45 kg (99 lb).

Ecological impacts

Grass carp (*Ctenopharyngodon idella*) have both positive and negative ecological impacts:

1. Positive Impacts

- **Aquatic Weed Control:** Grass carp play a vital role in managing excessive aquatic plant growth. Their herbivorous diet helps control aquatic vegetation, which can otherwise become problematic in lakes and rivers.
- **Water Quality Improvement:** By consuming large quantities of aquatic plants, grass carp indirectly improve water quality. Reduced plant biomass can lead to clearer water and better oxygen levels.

2. Negative Impacts

- **Biodiversity Reduction:** Grass carp can significantly alter ecosystems by reducing the diversity and abundance of aquatic plants. This can affect other organisms that rely on these plants for habitat and food.
- **Cascading Effects:** The decline in aquatic plants due to grass carp can have cascading effects on insects, fish, and birds that depend on these plants for survival.

Silver carp

Habitat Preferences and Distribution

Silver carp (*Hypophthalmichthys molitrix*) are one of four non-native fish species commonly referred to as invasive carp. Silver carp are native to temperate waters in eastern Asia. They can be found in river systems such as the Yangtze, West River, Pearl River, Kwangsi, Kwangtung, and Liao River in southern and central China. They also inhabit the Amur River basin in Russia, including rivers like the Amur, Amglluy, Arguan, Kerulean, Sungari (Songhuajiang), and Seya.

Preferred Habitat: Silver carp prefer large, open water habitats with abundant algae and plankton for food. They are often found in rivers and lakes with high nutrient levels. Outside of the United States, silver carp are commonly found in Asia, particularly in countries like China, Vietnam, and Thailand. They can tolerate a broad range of water temperatures, from 4°C to 40°C. Silver carp are known to feed at water temperatures of 10 to 19°C.

Morphological Characters

Silver carp have a narrow, deep body that is laterally compressed. Their body is covered in very small silver scales, except for their head, which is scaleless. The head and opercles (gill covers) lack scales. They fade from a greenish color on the back to silver on the belly as they mature. They have a large, toothless mouth with an upturned lower jaw that lacks barbs. Their eyes are situated far forward on the head, below the mouth, and project downwards. Silver carp can grow upwards of 3 feet long and commonly reach 20 pounds, with the largest individuals reaching upwards of 80 pounds. They are popularly known for their acrobatic jumps out of the water when startled by boat motors, becoming potential hazards for recreational boaters.

Feeding Adaptations and Dietary Habits

Silver carp are filter feeders, which means they primarily consume tiny particles suspended in the water. They possess a specialized feeding apparatus called gill rakers. These gill rakers are fused into a sponge-like filter, allowing them to capture particles as small as 4 μm. An epibranchial organ secretes mucus, aiding in trapping small food particles. A strong buccal pump forces water through this filter system. Silver carp lack stomachs, leading researchers to believe they feed more or less constantly. Their diet primarily consists of phytoplankton, which are microscopic plants that float in the water. They also consume zooplankton and detritus. In areas where silver carp have been introduced, they may compete with native planktivorous fish species (Hayer *et al.*, 2014).

Reproduction and Early Life Stages

Like many other fish species, silver carp reproduce through external fertilization. During the spawning season (typically late spring or early summer), males release sperm into the water, and females release their eggs. The fertilized eggs then develop into larvae, which eventually mature into juveniles. Silver carp eggs and larvae drift downstream in their natural range. Young fish hatch in the floodplain zone after drifting downstream. The larvae have a short period of development while drifting, and then they must swim from the river to find appropriate low- or no-flow nursery areas (Dong and Li, 1994).

Growth Dynamics and Lifespan

Silver carp grow rapidly during their early years and reach their maximum size relatively quickly. As they age, their growth rate decreases, and their size tends to plateau. Environmental factors such as temperature and food availability also play a significant role in determining the growth and size of silver carp. In their native range, silver carp reach maturity between 4 and 8 years old. However, in North America, they can mature as early as just 2 years old. Silver carp can live up to 20 years.

Ecological Impacts

Negative Impacts

Competition: Silver carp are efficient foragers and can outcompete native fish for food and habitat. They consume large quantities of phytoplankton, which reduces the available food for other aquatic species.

Habitat Destruction: Their voracious appetite for aquatic vegetation can lead to the destruction of important habitats for native fish and other aquatic organisms.

Jumping Behavior: Silver carp are known for their acrobatic jumps out of the water when startled by boat motors. Unfortunately, this behavior can harm boaters and damage boats.

Positive Impacts

Water Quality Control: Silver carp were introduced to some regions (including the United States) for aquaculture and recreational fishing purposes. They were initially used as biological control agents to reduce algae growth and improve water quality in ponds and wastewater treatment.

Bighead Carp

The bighead carp (*Hypophthalmichthys nobilis*) is a cyprinid freshwater fish native to East Asia. Here are some key points about its habitat and distribution:

Habitat: Bighead carp inhabit lakes, rivers, and reservoirs. They prefer high fertility water with abundant natural food. These fish normally dwell in the upper layer of the water column. Successful reproduction requires a riverine environment with long reaches, high water volume, turbulent flow, and warm water temperatures (66 to 84 °F).

Distribution: Native to China, their range extends from southern China north to the Amur River system, which forms the border between China and Russia. They have been introduced widely outside their native range, including in the United States, where they are considered invasive due to out-competing native species.

Morphological Characters

The bighead carp (*Hypophthalmichthys nobilis*) exhibits the following morphological characteristics: Large and deep-bodied fish, with a narrow profile. Scaleless, with eyes positioned far forward and low on the head, projecting downward. Dark gray body fading to white on the underside, with irregular dark blotches on the side. Large and toothless, with a protruding lower jaw. Typically around 60 cm (2 ft), but can reach up to 146 cm (4 ft 9 in) and 40 kg (88 lb).

Feeding Adaptations and Dietary Habits

The bighead carp (*Hypophthalmichthys nobilis*) predominantly engages in filter feeding. While it prioritizes zooplankton consumption, it also feeds on phytoplankton and detritus, showcasing its adaptability and versatile dietary preferences. These carp are powerful filter-feeders with a wide food spectrum, growing fast and reproducing quickly. They strain plankton and detritus from the water using their closely set gill rakers (Schrank and Guy, 2002).

Reproduction and Early Life Stages

Bighead carp spawn in turbulent rivers. Their eggs develop and hatch while drifting downriver. The larvae also spend a short period developing in the drift. Afterward, they must swim from the river and find appropriate low- or no-flow nursery areas. A male becomes sexually mature after about 2 years, while females often require 3 years of age before they can reproduce. The eggs and early larvae of bighead carp drift in the current. They are denser than water and are thought to perish if the current and turbulence are insufficient to keep them from settling to the bottom before they reach a stage at which they can swim (Jennings, 1988).

Growth Dynamics and Lifespan

Bighead carp reach full maturity in 2 to 3 years. Their growth rate is influenced by factors such as temperature, food availability, and habitat conditions. Earlier maturation results in greater subadult survival, but lower adult survival and fecundity. Bighead carp can live for at least 9 years. Their lifespan, combined with rapid proliferation, allows them to potentially dominate ecosystems within a short time.

Ecological Impacts

Negative Impacts

Ecosystem Disruption: Bighead carp are considered highly invasive in the river watershed and lakes. They alter natural habitats, disrupting ecosystems.

Competition: They consume plankton and vegetation, which are crucial for native fish species. This competition affects the food web and fisheries.

Contaminants: Bighead carp may contain high levels of contaminants, posing potential hazards to human health.

Positive Impact (Limited)

In some cases, piscivorous fish like walleye (*Sander vitreus*) may benefit from bighead carp establishment.

Conclusion

The introduction of non-native species, both accidental and intentional, has significant ecological implications. Invasive species, as defined by the IUCN, establish themselves in new environments, alter ecosystems, and threaten biodiversity. Various reasons, such as enhancing local fisheries, promoting sport fishing, and controlling pests, have driven the introduction of exotic fish species worldwide. However, these introductions often lead to the extinction of native species, competition for resources, and the spread of diseases. Aquaculture has been a primary motivator for moving fish across regions, aiming to boost productivity and meet food security demands. This practice has introduced species like common carp, grass carp, and silver carp into India, often benefiting economic interests but also posing ecological risks. Invasive species can disrupt ecosystems, outcompete native species, and cause long-term environmental and socioeconomic harm. Effective management and monitoring of introduced species are essential to balance economic benefits with ecological preservation, ensuring the protection of native biodiversity and the health of aquatic ecosystems.

References

Biju Kumar, A. (2000). Exotic fishes and freshwater diversity, Zoos Print Journal 15 (11): 363-367.

Dong, S. and Li, D. (1994). Comparative studies on the feeding selectivity of silver carp Hypophthalmichthys molitrix and bighead carp Aristichthys nobilis. Journal of Fish Biology, 44(4), 621-626.

Hayer, C. A., Breeggemann, J. J., Klumb, R. A., Graeb, B. D., & Bertrand, K. N. (2014). Population characteristics of Bighead and Silver Carp on the northwestern front of their North American invasion. Aquatic Invasions, 9(3).

Jennings, D. P. (1988). Bighead carp (Hypophthalmichthys nobilis): biological synopsis (Vol. 88, No. 29). Fish and Wildlife Service, US Department of the Interior.

Joshi, K. D., Basheer, V. S., Kumar, A., Srivastava, S. M., Sahu, V. and Lal, K. K. (2021). Alien fish species in open waters of India: Appearance, establishment and impacts. The Indian Journal of Animal Sciences, 91(3), 167-173.

Kottelat, M. and T. Whitten. 1996. Freshwater biodiversity in the Asia with special reference to fish. World Bank Technical Paper No. 343, Washington, 59 pp

Li, L., Balto, G., Xu, X., Shen, Y. and Li, J. (2023). The feeding ecology of grass carp: a review. Reviews in Aquaculture, 15(4), 1335-1354.

Mohd, I., Bhat, F. A., Shah, T. H., Wali, A. and Nabi, I. (2023). Reproductive characteristics of wild and captive Scale carp, Cyprinus carpio var. communis in Kashmir, India.

Moyle, P. B. and R. A. Leidy 1992. Loss of biodiversity in aquatic ecosystems: evidence from fish faunas, In: Fiedler, P.L. and S.K. Jain (eds.), Conservation Biology: The Theory and Practice of Nature Conservation, Preservation and Management. Chapman and Hall, New York, pp. 127-169.

Nyman, L. 1991. Conservation of freshwater fish. Protection of biodiversity and genetic variability in aquatic ecosystems. Fisheries Development Series 56, Swedmar & WWF, Sweden, 38 p.

Philipp, D.P., J.M. Epifanio, J.E. Mardsen and J.E. Claussen (eds.) 1995. Protection of Aquatic Biodiversity. Proceedings of the World Fisheries Congress. Theme 3, Oxford and IBH Publ. Co., New Delhi, 282 p.

Raghubanshi, A.S., L.C. Rai, J.P. Gaur and J.S. Singh .2005. Invasive alien species and biodiversity in India, Current Science, 88(4): 539-540

Raman, R. P., Mishra, A., Kumar, S., Sahay, S., Bhagat, M. N. and Kumar, S. (2013). Introductions of exotic fish species into Indian waters: An overview of benefits, impacts. Advances in fish research, 6, 1-14.

Schrank, S. J. and Guy, C. S. (2002). Age, growth, and gonadal characteristics of adult bighead carp, Hypophthalmichthys nobilis, in the lower Missouri River. Environmental Biology of fishes, 64, 443-450.

Singh, A. K., Pathak, A. K. and Lakra, W. S. (2010). Invasion of an exotic fish—common carp, Cyprinus carpio L. (Actinopterygii: Cypriniformes: Cyprinidae) in the Ganga River, India and its impacts. Acta Ichthyologica et Piscatoria, 40(1), 11-19.

Swain, S., Felix, S., Anthony, C. and Chhandaprajnadarsini, E. M. (2017). Exotic fish species introduced in India and its impacts. Scientific India, 27-29.

Welcome, R .L. 1988. International introduction of inland aquatic species. FAO Fish Tech. 294-318.

Zhao, Y., Zhang, L., Wang, C. and Xie, C. (2020). Biology and ecology of grass carp in China: a review and synthesis. North American Journal of Fisheries Management, 40(6), 1379-1399.

8

Morphological Features, Habitat Preferences, and Ecological Role of Loaches in Kashmir Valley

Mir Iqra Farooq*, Tasaduq Hussain Shah, Hudisa Banoo Arizo Jan and Inab Majeed Bala

Division of Fisheries Resource Management, Faculty of Fisheries, SKUAST-K Rangil Ganderbal, Kashmir, Jammu and Kashmir

Abstract

The Kashmir Valley, located in the western Himalayas, is a region rich in water resources and diverse aquatic life. This chapter explores the ecosystem of the valley, emphasizing the significance of its water bodies, which cover more than 6% of its land surface. The valley's continuous supply of water from glaciers, streams, and springs supports a variety of fish species, particularly from the Cypriniformes Order. Notable among these are the loaches, including species such as Triplophysa kashmirensis, T. marmorata, Paraschistura punjabensis, and Botia birdii. These species exhibit unique morphological, ecological, and behavioral adaptations that enable them to thrive in the fast-flowing, cold-water habitats of the region. The chapter details their feeding ecology, reproductive strategies, growth rates, and longevity, highlighting their roles in nutrient cycling and as bioindicators for freshwater ecosystem health. The conservation status of these species is also discussed, with some, like T. kashmirensis, being critically endangered.

***Keywords:** Loaches, Cypriniformes, Kashmir Valley, Cold-water, Freshwater ecosystem, Critically endangered*

Introduction

The Kashmir Valley, situated in the western Himalayas between latitudes 33°01'N and 35°00'N and longitudes 73°48'E and 75°30'E, spans an area of 4,920 square kilometers and sits at an elevation of 1,585 meters. This region is rich in water resources, featuring numerous lakes, rivers, and wetlands. The ecosystem benefits from a continuous supply of water from perennial glaciers, mountain streams, and a multitude of springs. It is estimated that more than

6% of Kashmir's land surface is covered by various types of water bodies, enhancing the valley's natural beauty (Zutshi and Gopal, 2000). Whether from glaciers, springs, rivers, or lakes, the water in Kashmir contributes significantly to its picturesque landscape. The Kashmir Valley hosts a variety of 23 fish species, both native and introduced (Bhat *et al.*, 2020). Cypriniformes, the largest group of freshwater fish, are divided into two main superfamilies: Cyprinoidea (carp like fishes) and Cobitoidea (loach fishes). This classification covers a wide variety of aquatic species, highlighting the extensive biodiversity within the Cypriniformes order (Nelson *et al*, 2016). The initial documentation of cold-water fish in India was provided by Hamilton in 1822. The Cobitidae family, known as true loaches, includes species such as ***Botia birdi*** found in the Kashmir Valley. The family Balitoridae, specifically the subfamily Nemacheilinae, in this region include two scaleless fish species earlier named as *Cobitis vittata* and *C. marmorata* by Heckel (1838) but Kullander *et al.* (1999) renamed them as *Triplophysa marmorata* and *T. kashmirensis* and ***Paraschistura punjabensis.*** Loaches are a fascinating subject for aquatic enthusiasts and researchers alike. Found predominantly in Eurasian waters, these bottom-dwelling fish are recognized for their distinctive elongated bodies, small scales, and barbels around the mouth which aid in their sensory and feeding functions. Loaches are renowned for their adaptability, thriving in a variety of habitats ranging from fast-flowing rivers to stagnant ponds. With over 200 species classified across several genera, loaches exhibit a remarkable range of behaviours and ecological roles, from the well-known clown loach (*Chromobotia macracanthus*), popular in the aquarium trade, to the stone loach (*Barbatula barbatula*), a species integral to the ecosystems of European rivers and streams (Nelson, 2006; Kottelat, 2012). Loaches play significant roles in their natural habitats, contributing to the benthic community dynamics and nutrient cycling. Their sensitivity to environmental changes makes them valuable bioindicators for assessing the health of freshwater ecosystems (Kottelat & Freyhof, 2007).

On the Qinghai-Tibet Plateau (QTP), species of the genus ***Triplophysa*** (Rendahl, 1933) are among the most prevalent Nemacheilids (Zhu, 1989; Wu and Wu, 1992). These fish are widespread across various water bodies in the region, and new species of ***Triplophysa*** continue to be discovered (Yang *et al.*, 2016; Huang *et al.*, 2016; Liu *et al.*, 2017; Wu *et al.*, 2018). To date, 160 valid ***Triplophysa*** species have been identified (Fricke *et al.*, 2020), with more than 85% of them found in China (He *et al.*, 2006; He *et al.*, 2011; Ren *et al.*, 2012; Froese and Pauly, 2013). While ***Triplophysa*** species are primarily distributed on the Qinghai-Tibet Plateau and surrounding areas (Zhu, 1989; Wu and Wu, 1991), they are also present in northern China and central Asia. The genus plays a significant role in the fish fauna of the Qinghai-Tibet Plateau. Reports indicate that this genus is found in the upper Indus River drainage in Jammu and Kashmir, as well as in the Lahul and Spiti regions of Himachal Pradesh (Kullander *et al.*, 1999). These fish inhabit rivers and streams in mountainous areas with rocky beds (Bashir *et al.*, 2016). Menon (1987) documented ten ***Triplophysa*** species in India.

Triplophysa is distinct from other genera due to its narrow nostrils, a dumbbell-shaped swim bladder, and a bony postero-lateral bladder sac. In males, six to eight branches of the pectoral fin become thicker, wider, and are covered with reproductive nodules. These nodules also appear on both sides of the head, extending from the eyes almost to the upper jaw bar insertion (Zhu, 1989; Yang, 1990). The species differentiation is high among isolated rivers and lakes. Due to their small size, these fish have limited interest in fisheries, leading to slower taxonomic progress compared to larger fish. The genus *Triplophysa* is widespread in Central Asia, ranging from Afghanistan and Balochistan to Upper Asia, including Lake Balkhash, Lake Urnur, and Outer Mongolia. They are also found in the Chinese provinces of Sichuan, Yunnan, Chongqing, Hunan, Guizhou, and Guangxi (Yang *et al.*, 2016). *Triplophysa* species exhibit notable sexual dimorphism: males have tubercle-bearing raised skin on the sides of the head and a thick tuberculate pad on the extended and widened pectoral-fin rays (Zhu, 1989). Though Barbatula species also show similar sexual dimorphism, *Triplophysa* can be distinguished by the close proximity of their nostrils (Banarescu and Nalbant, 1968; Prokofiev, 2010; Yang *et al.*, 2012; Liu *et al.*, 2017)

Loach Species Found in Kashmir

Triplophysa kashmirensis **(Hora, 1922)**

Classification

Kingdom	:	Animalia
Phylum	:	Chordata
Class	:	Actinopterygii (ray-finned fishes)
Order	:	Cypriniformes
Family	:	Nemacheilidae (stone loaches)
Genus	:	Triplophysa
Species	:	*Triplophysa kashmirensis*

Local name: Ara Gurun

Common name: Kashmir loach

Characteristics: This species exhibits several distinct morphological features that enable it to thrive in its specific habitat: **it has** typically an elongated, slightly compressed body which is well-suited for living in fast-flowing waters. The streamlined shape helps to reduce resistance and allows for efficient movement against strong currents. The average length of adult ***T. kashmirensis*** ranges from 10 to 15 cm, although some individuals may grow larger depending on environmental conditions (Kullander *et al.*, 1999). The head is relatively small with a slightly pointed snout. The species is equipped with three pairs of barbels (one maxillary and two rostral), to detect food particles in the substrate, employing a combination of tactile and chemical cues. The dorsal fin is positioned midway along the body and is composed of soft rays, which provide stability and manoeuvrability in turbulent waters. Pectoral fins are broad and strong, aiding in maintaining position and navigating through the riverbed. The caudal fin is slightly forked, which enhances propulsion (Kullander *et al.*, 1999). It exhibits a brown to olive coloration with dark mottling or spots along its body, providing camouflage against the rocky and sandy riverbeds. The ventral side is generally lighter, aiding in blending with the water surface when viewed from below.

Habitat preferences and distribution: The species prefers habitats with a gravel or rocky substrate, where it can forage for food and find shelter from predators and strong currents (Kullander *et al.*, 1999). The distribution of *T. kashmirensis* is largely confined to the high-altitude regions of South Asia, particularly within the Himalayan mountain range. This species is predominantly found in the Kashmir Valley of India, including the Jhelum River, Dal Lake, and Wular Lake, as well as in the high-altitude streams of the Ladakh region, particularly the upper reaches of the Indus River and its tributaries (Gupta and Banerjee, 2013; Joshi and Singh, 2015). It is also present in similar high-altitude, cold-water environments in the northern areas of Pakistan, specifically regions contiguous with Indian Kashmir (Kullander and Fang, 2004).

Conservation status: Critically Endangered (IUCN, 2023)

2. Triplophysa marmorata **(Heckel, 1838)**

Classification

Kingdom	:	Animalia
Phylum	:	Chordata
Class	:	Actinopterygii (ray-finned fishes)
Order	:	Cypriniformes
Family	:	Nemacheilidae (stone loaches)
Genus	:	Triplophysa
Species	:	*Triplophysa marmorata*

Local name: Ara Gurun

Common name: Marble loach

Characteristics: It has an elongated, somewhat cylindrical body that tapers towards the tail. This streamlined shape aids in reducing water resistance, facilitating easier movement in swift currents. The standard length of adults typically ranges from 8 to 15 cm, although some individuals may grow larger under favourable conditions. The head of *T. marmorata* is relatively small and slightly flattened. The mouth is subterminal and positioned to assist in bottom-feeding. The presence of barbels around the mouth is a notable feature; these sensory organs help the fish detect food particles in the substrate. The species typically has three pairs of barbels one maxillary and two rostral. The pectoral fins are broad and well-developed, providing stability and maneuverability in turbulent waters. The dorsal fin is situated approximately mid-body and is composed of a single short base followed by a long, flexible fin that assists in balance and propulsion. The pelvic fins are also relatively large, positioned slightly posterior to the pectoral fins, enhancing the fish's ability to anchor itself against the substrate (Prokofiev, 2010). The body coloration and patterns help it camouflage against the rocky substrates of its habitat. The caudal peduncle is slender, leading to a slightly forked caudal fin. This fin structure is efficient for sudden bursts of speed, which can be crucial for escaping predators and navigating through the complex substrates of their habitats.

Habitat preferences and distribution: *T. marmorata* thrives in cold, oxygen-rich, fast-flowing waters with rocky substrates, which provide ideal conditions for its benthic lifestyle (Wu and Wu, 1992). *T. marmorata* is primarily distributed across high-altitude freshwater systems in Asia. It is predominantly found in the Tibetan Plateau, including Tibet Autonomous Region and Qinghai Province in China, as well as in high-altitude rivers of Yunnan Province (Chen *et al.*, 2012). The species also inhabits the Kashmir and Ladakh region of India, northern areas of Pakistan such as Gilgit-Baltistan, and various Himalayan streams in Nepal (Mirza, 1991; Shrestha, 2008; Joshi and Singh, 2015)

Conservation status: Least concern (IUCN, 2023)

3. Paraschistura punjabensis (Hora, 1923)

Classification

Kingdom : Animalia

Phylum : Chordata

Class : Actinopterygii

Order : Cypriniformes

Family : Nemacheilidae

Genus : Paraschistura

Species : *Paraschistura punjabensis*

Local name: Ara Gurun

Common name: Punjab Loach

Characteristics: It has an elongated and somewhat compressed body, which is typical of benthic fish that inhabit fast-flowing streams. This body shape helps it navigate and maintain position in turbulent waters (Kottelat, 1990). Adults typically range from 6 to 8 cm in length, though size can vary depending on environmental conditions. The head is moderately sized with a slightly pointed snout. The species has three pairs of barbels (one maxillary and two rostral), which are sensory organs used to detect food in the substrate. These barbels are an essential adaptation for their benthic lifestyle. The dorsal fin is located midway along the body and consists of soft rays, providing stability in fast-flowing water. Pectoral fins are broad and help the fish maintain its position on the riverbed. The caudal fin is slightly forked, aiding in propulsion and manoeuvrability (Kottelat, 1990). The coloration of *P. punjabensis* typically includes a series of dark vertical bars against a lighter background, providing effective camouflage against the stony riverbeds. This pattern helps the fish blend into its surroundings, reducing the risk of predation (Kullander *et al.*, 1999).

Habitat preferences and distribution: This species is found in clear, fast-flowing streams and rivers with a rocky or gravel substrate. It prefers habitats with ample cover and moderate to high current velocities (Kottelat, 1990). This species is endemic to the Indus River system in the Punjab region, in both Pakistan and northern India including Kashmir region.

Conservation status: vulnerable (IUCN, 2023)

4. *Botia birdii* (Chaudhuri, 1909)

Classification

Kingdom : Animalia

Phylum : Chordata

Class : Actinopterygii

Order : Cypriniformes

Family : Botiidae

Genus : Botia

Species : *Botia birdii*

Local name: Ram gurun

Common name: Bird Loach

Characteristics: It has an elongated and slightly compressed body, which is typical of loaches, allowing it to navigate and dwell in the benthic zone of rivers and streams (Talwar and Jhingran, 1991). Adult individuals typically reach lengths of 10 to 15 cm, though sizes can vary depending on environmental conditions. The head is relatively short with a blunt snout. It has three pairs of barbels around the mouth, which are sensory organs used to detect food in the substrate. These barbels are a significant adaptation for their bottom-dwelling lifestyle (Menon, 1999). The dorsal fin is situated midway along the body and has a combination of spines and soft rays, providing stability and manoeuvrability in flowing waters. Pectoral fins are broad and help the fish maintain position on the riverbed. The caudal fin is slightly forked, aiding in propulsion and efficient swimming (Talwar and Jhingran, 1991). **The species** typically exhibits a yellowish to brown coloration with dark vertical bands or stripes along its body, which help in camouflaging against the rocky and sandy riverbeds. The ventral side is lighter, aiding in blending with the water surface when viewed from below (Menon, 1999).

Habitat preferences and distribution: This species is found in clear, fast-flowing streams and rivers with a rocky or gravel substrate. It prefers habitats with abundant cover and moderate to high current velocities (Talwar and Jhingran, 1991). It is primarily distributed in South Asia, with significant populations in India, Pakistan, and Nepal. It is found in the freshwater systems of Kashmir, such as the Jhelum River, Dal Lake, and Wular Lake (Talwar and Jhingran, 1991; Jayaram, 2010). In Pakistan, it is present in the Indus River system, and in Nepal, it resides in the river systems of the southern plains (Wu, 2012).

Conservation status: Least concern (IUCN, 2023)

Feeding Ecology and Trophic Roles

Loaches exhibit diverse feeding habits and trophic roles within their ecosystems. Loaches are primarily benthic feeders, meaning they forage on or near the bottom of water bodies. Their diet and feeding strategies vary depending on species, habitat, and availability of food resources. Many loaches feed on a variety of benthic invertebrates such as insect larvae (e.g., Ephemeroptera, Trichoptera, Diptera), small crustaceans, and mollusks. This diet provides them with essential nutrients and energy for growth and reproduction (Kottelat, 1990). Some loach species, such as those in the genus ***Triplophysa***, also consume detritus and periphytic algae, which are critical for their survival in nutrient-poor environments (Yang and Cao, 2001). Larger loaches, like those in the genus **Botia,** may occasionally feed on small fish and fish eggs, especially in environments where these resources are readily available (Talwar and Jhingran, 1991). Loaches use their well-developed barbels to detect food in the substrate. The barbels are sensitive to touch and chemical cues, allowing loaches to locate prey hidden within the sediment (Menon, 1999). Some loach species are primarily diurnal, others exhibit nocturnal feeding habits, adapting their activity patterns to reduce competition and predation (Kottelat, 1990). The diet of loaches can vary seasonally, with increased consumption of certain prey items like insect larvae during their peak emergence periods. This seasonal dietary shift ensures that loaches exploit the most abundant and nutritious food resources available at different times of the year (Du and Chen, 2010).

Loaches play significant trophic roles in freshwater ecosystems, influencing both lower and higher trophic levels through their feeding activities. As mid-level predators, loaches help control populations of benthic invertebrates. This predation pressure prevents any single species from becoming overly dominant, thereby maintaining the ecological balance and promoting biodiversity (Kottelat, 1990). By feeding on detritus and algae, loaches contribute to nutrient cycling within their ecosystems. Their foraging activities help break down organic matter, releasing nutrients back into the water column and supporting primary production (Yang and Cao, 2001). Loaches serve as a food for larger fish, birds, and other predators. This role makes them integral components of the food web, facilitating energy transfer from lower to higher trophic levels (Talwar and Jhingran, 1991). The presence of loaches can influence the distribution and behavior of their predators, shaping the overall structure of the aquatic community (Menon, 1999). Through their feeding and foraging activities, loaches contribute to the health and stability of freshwater ecosystems. Their role in regulating invertebrate populations and recycling nutrients supports a dynamic and resilient aquatic environment (Du and Chen, 2010).

Reproductive Strategies and Spawning Behaviours

Loaches have developed specific reproductive strategies and spawning behaviours that ensure the survival and dispersal of their offspring in challenging environments. ***T. kashmirensis* and *T. marmorata*** spawns in the late spring to early summer, typically from May to July, when the water temperature rises and conditions become more favourable for egg development and larval survival (Raina and Jamila, 1996; Wang *et al.*, 2001). Females produce a large number of eggs to increase the chances of survival. The high fecundity compensates for the high predation rates and harsh environmental conditions. ***T. kashmirensis*** lays its eggs on gravel or rocky substrates in fast-flowing streams. This choice of substrate ensures that the eggs receive sufficient oxygen and are protected from being swept away by the current (Raina and Jamila, 1996). The eggs of ***T. kashmirensis* and *T. marmorata*** are adhesive, allowing them to stick to the substrate and remain in place until they hatch. This adaptation is crucial for maintaining the eggs' position in fast-flowing waters (Raina and Jamila, 1996). ***T. marmorata*** is known to release eggs in several batches over the spawning season. This strategy reduces the risk of losing all offspring to sudden environmental changes or predation events.

B. birdii typically spawn during the monsoon season. The increased water levels and temperature changes act as triggers for spawning, which aligns their reproductive cycle with environmental conditions conducive to the survival of their offspring. This species exhibits batch spawning behaviour, where females release eggs in batches over a period rather than all at once. This strategy increases the chances of fertilization and subsequent survival of the offspring in varying environmental conditions (Sarkar *et al.*, 2016). Changes in water temperature and flow, as well as the availability of suitable substrates, play critical roles in triggering spawning. These environmental cues ensure that the eggs and larvae have optimal conditions for development (Kumar *et al.*, 2015). During the spawning season, *B. birdii* exhibit distinct courtship behaviours. Males become more vibrant in color and engage in elaborate displays to attract females. These displays often involve physical posturing and gentle nudging (Siddiqui, 2018). *B. birdii* are substrate spawners, preferring to lay their eggs on surfaces such as rocks, gravel, or aquatic plants. This behaviour protects the eggs from predators and provides a stable environment for development. Females lay adhesive eggs that stick to the substrate. Males then fertilize the eggs externally. The adhesive nature of the eggs prevents them from being swept away by currents, ensuring they remain in a safe location (Singh and Saini, 2016). While *B. birdii* do not exhibit extensive parental care post-spawning, their selection of secure spawning sites and the adhesive nature of the eggs provide initial protection for the developing embryos (Joshi and Chandra, 2020).

P. punjabensis typically spawns during the pre-monsoon and monsoon seasons. The onset of rains and the associated changes in water parameters, such as increased flow and temperature, act as natural triggers for the onset of spawning (Shrestha *et al.*, 2017). This species is iteroparous, meaning it spawns multiple times over its lifespan. This strategy allows the fish to take advantage of favorable conditions at different times, increasing the likelihood of successful reproduction. Factors such as water temperature, photoperiod, and water chemistry play crucial roles in the reproductive cycle of *P. punjabensis*. These environmental cues ensure synchronization of spawning with optimal conditions for larval survival. Males of *P. punjabensis* exhibit courtship behaviours to attract females. These behaviours include vibrant coloration changes, fin displays, and chasing. Such behaviours are crucial for mate selection and successful spawning (Kumar and Tandon, 2016). *P. punjabensis* is a substrate spawner. It prefers to lay its eggs on gravel or rocky substrates in streams with moderate to fast flow. This choice of substrate helps in protecting the eggs from predation and strong currents. Females lay adhesive eggs that stick to the chosen substrate. Males follow to externally fertilize the eggs. The adhesive nature of the eggs prevents them from being washed away, ensuring they remain in place until hatching (Singh and Singh, 2020). Post-spawning, *P. punjabensis* exhibits minimal parental care. The eggs are left to develop on their own, relying on the protective substrate and favorable environmental conditions selected by the parents during spawning.

Growth Rates and Longevity

Loaches relatively have long lifespan which is typical for cold-water fish, which often have slower metabolisms and longer developmental periods (Wang and Li, 2013; Ahmad *et al.*, 2017). Stable, clean water conditions and a consistent and adequate supply of natural food sources is essential for maintaining their health and longevity. Lower predation pressure and reduced competition for food resources in their natural habitats contribute to their longevity (Sarkar *et al.*, 2017). Seasonal fluctuations in food availability, any change in water quality, temperature, or habitat structure can adversely affect their lifespan can influence their overall lifespan (Rishi and Jain, 2012; Mir *et al.*, 2013; Raina *et al.*, 2015; Xie *et al.*, 2016; Chandra *et al.*, 2017). *T. kashmirensis* exhibits relatively slow growth rates typical of cold-water fish. Juveniles grow approximately 1-2 cm per year in the early stages of life. This slower growth rate is a result of the cold water temperatures and limited food resources in their natural habitat (Mir *et al.*, 2013). As the fish mature, the growth rate slows further. Adults typically reach an average length of 10-15 cm over a period of several years (Ahmad *et al.*, 2017). Cold water temperatures in the streams of the Kashmir Valley significantly slow down the metabolic rate, thereby affecting the overall growth rate of *T. kashmirensis* (Raina *et al.*, 2015). The availability of natural food sources such as small invertebrates and detritus plays a crucial role in the growth of these loaches. Seasonal variations in food availability can also impact their growth rates (Mir *et al.*, 2013). *T. kashmirensis* is known to have a lifespan of around 5-8 years in the wild.

Juvenile *T. marmorata* grow relatively quickly during their first year, typically reaching a length of 5-7 cm. This rapid early growth is crucial for survival, as it helps them avoid predation and adapt to their environments (Zhao *et al.*, 2015). After the initial rapid growth phase, the growth rate slows down. By the second year, they may reach around 10-12 cm, and growth continues more gradually thereafter, influenced by factors such as water temperature, food availability, and habitat conditions (Chen and Yu, 2014). Being a cold-water species, *T. marmorata* growth rate is closely tied to water temperature. Optimal growth occurs in cooler, well-oxygenated waters, typical of high-altitude streams and rivers (Xie *et al.*, 2016). The availability and quality of food sources significantly affect their growth rate. Seasonal variations in food availability can lead to fluctuations in growth (Zhao *et al.*, 2015). *T. marmorata* can live for about 8-10 years in the wild.

Juveniles of *B. birdii* exhibit moderate growth rates, typically reaching about 4-6 cm within their first year. This growth rate can be influenced by the availability of food and optimal water conditions (Sarkar *et al.*, 2017). As *B. birdii* matures, its growth rate slows down. Adults generally attain a size of 15-20 cm over several years. Growth beyond this size is gradual and can be affected by environmental factors such as water quality, temperature, and food supply (Das and Ray, 2018). Growth rates are higher in warmer waters, which enhance metabolic activity and food assimilation. However, extreme temperatures can stress the fish and reduce growth rates (Roy and Dey, 2015). A diet rich in protein, consisting of live or frozen foods like worms, insects, and small crustaceans, promotes better growth. Poor nutrition or limited food availability can stunt growth. *B. birdii* has a relatively long lifespan for a freshwater fish, typically living 8-12 years in the wild. In captivity, with optimal care, they can live up to 15 years or more (Gupta and Banerjee, 2019). Juvenile *P. punjabensis* exhibit relatively rapid growth in their first year, typically reaching about 3-4 cm in length. This initial growth rate is influenced by the availability of food and optimal water conditions (Rishi and Jain, 2012). As they mature, their growth rate slows. By the second year, individuals may reach around 5-7 cm, and adults generally attain a size of 8-10 cm over several years. Growth rates can vary based on environmental factors and habitat conditions (Kumar and Tandon, 2014). Optimal growth occurs in clean, well-oxygenated, and moderately flowing waters. Temperatures between 18-25 °C are ideal for their growth. Extreme temperatures and poor water quality can inhibit growth (Singh *et al.*, 2016). A varied diet rich in small invertebrates, detritus, and algae promotes better growth. Limited food availability or poor nutrition can negatively impact growth rates (Sharma and Joshi, 2015). *P. punjabensis* typically has a lifespan of about 5-7 years in the wild.

Threats

Agricultural runoff, untreated sewage, and industrial discharges: introduce pollutants such as pesticides, fertilizers, heavy metals, and other chemicals into the water bodies where these loaches live. This pollution degrades water quality, affecting the health and survival of the fish (Raina *et al.*, 2015; Roy and Dey, 2015).

Sedimentation: Deforestation and land-use changes lead to increased sediment load in rivers and streams. Sedimentation can smother eggs, reduce feeding efficiency, and alter habitats (Kumar and Tandon, 2014).

Dam Construction: Dams and other water diversion projects alter the flow regimes of rivers, which can disrupt spawning cycles, reduce habitat connectivity, and change water temperature and oxygen levels. These changes adversely affect the life cycles of loach species (Singh *et al.*, 2016; Sarkar *et al.*, 2017).

Water Extraction: Excessive withdrawal of water for irrigation, drinking, and industrial purposes lowers water levels, which can reduce habitat availability and quality (Chandra *et al.*, 2017).

Overfishing: Overfishing for local consumption can lead to population declines. This is particularly problematic when fish are caught before they have had a chance to reproduce, leading to a decrease in population size over time (Ahmad *et al.*, 2017).

Aquarium Trade: Some species, such as *B. birdii,* are also collected for the aquarium trade. Overharvesting for this purpose can significantly reduce wild populations (Gupta and Banerjee, 2019).

Competition and Predation: The introduction of non-native species can lead to competition for resources, predation on native loaches, and the spread of diseases. Invasive species can outcompete native loaches for food and habitat, leading to population declines (Sharma and Joshi, 2015).

Climate Change: Climate change can lead to alterations in water temperature, which can affect the metabolic rates, growth, and reproduction of these species. Warmer temperatures can also increase the prevalence of diseases and parasites (Das and Ray, 2018).

Hydrological Changes: Changes in precipitation patterns can alter river flows, potentially leading to more frequent and severe droughts or floods, both of which can negatively impact loach habitats (Singh *et al.*, 2016).

Conclusion

The rich aquatic ecosystems of the Kashmir Valley, characterized by its diverse array of loach species, play a vital role in maintaining the ecological balance of freshwater habitats. From the fast-flowing streams of the Jhelum River to the serene waters of lakes like Wular, these environments host a remarkable variety of loaches, each adapted to its specific niche within the ecosystem. *T. kashmirensis, T. marmorata*, and *B. birdii* are three of the more well-known loach species that are found in the Kashmir region. These species differ from one another in terms of their physical characteristics, eating preferences, and habitat-specific reproductive tactics. These species play a major role in the trophic structure of their environments by acting as both predators and prey and by transferring energy and nutrients within freshwater food webs. However, despite their ecological importance, these loach populations face numerous threats, including habitat degradation, pollution, overfishing, and the impacts of climate change. Anthropogenic activities such as agricultural runoff, dam construction, and water extraction pose significant challenges to the conservation of these species and their habitats. Comprehensive conservation methods that address the direct and indirect sources of population decline are necessary to offset these challenges and preserve loach populations. The biodiversity and ecological integrity of the freshwater ecosystems in the Kashmir Valley depend on conservation efforts such habitat restoration, pollution prevention, sustainable fisheries management, and the creation of protected areas. Furthermore, research efforts aimed at better understanding the biology, ecology, and population dynamics of loach species will provide valuable insights into their conservation needs and inform evidence-based management decisions. Collaborative initiatives involving scientists, policymakers, local communities, and conservation organizations are crucial for achieving long-term conservation goals and ensuring the survival of these iconic freshwater species for future generations.

References

Ahmad, R., Mir, F.A., & Yousuf, A. R. (2017). Reproductive Biology and Longevity of Triplophysa kashmirensis in the Streams of Kashmir Valley. Environmental Biology of Fishes, 100(5), 635-644.

Banarescu, P. M., & Nalbant, T. T. (1968). Revision of the genus Barbatula (Pisces, Cobitidae). Vestník Ceskoslovenske Spolecnosti Zoologicke, 32(1), 9-37.

Bashir, I., Yousuf, A. R., & Parveen, M. (2016). Fish diversity of Wular Lake, Kashmir Himalayas, and their impact on the fishery of the lake. Journal of Threatened Taxa, 8(3), 8602-8609.

Bhat, F.A., Yousuf, A.R., Balkhi, M.H. (2020). Diversity of Fishes in Jammu and Kashmir State. In: Dar, G., Khuroo, A. (eds) Biodiversity of the Himalaya: Jammu and Kashmir State. Topics in Biodiversity and Conservation, vol 18. Springer, Singapore.

Chandra, K., Singh, A. K., & Pathak, J. K. (2017). Effect of Environmental Parameters on the Longevity of Freshwater Fish Species Including Botia birdii. Journal of Environmental Science and Health, 52(9), 1356-1364.

Chen, X., & Yu, D. (2014). Environmental Influences on the Growth Rate of Triplophysa marmorata in High-Altitude Streams. Environmental Biology of Fishes, 97(4), 453-462.

Das, S. K., & Ray, S. (2018). Studies on the Growth Patterns of Botia birdii in Freshwater Systems of India. Aquaculture Research, 49(6), 2022-2030.

Du, L.N., & Chen, X.Y. (2010). Feeding habits of Triplophysa bleekeri in the headwaters of the Yangtze River, China. Environmental Biology of Fishes, 87(1), 39-47.

Fricke, R., Eschmeyer, W. N., & Fong, J. D. (2020). Species of Fishes by Family/Subfamily. Eschmeyer's Catalog of Fishes: Genera, Species, References

Froese, R., & Pauly, D. (Eds.) (2011). FishBase. World Wide Web electronic publication

Froese, R., & Pauly, D. (Eds.) (2013). FishBase. World Wide Web electronic publication

Gupta, S., & Banerjee, S. (2019). Longevity and Health Management in Botia birdii in Captivity. International Journal of Aquatic Sciences, 7(1), 112-120.

Gupta, S., & Banerjee, S. (2013). Fish and fisheries of high altitude lakes in India." Fishery Technology, 50(1), 51-56

He, D., Chen, Y., & Chen, Y. (2011). "Distribution and conservation status of endemic fishes in the upper Yangtze River, China." Aquatic Conservation: Marine and Freshwater Ecosystems, 21(6), 542-554.

Huang, W., Liu, Q., Xie, J., Wang, W., Xiao, J., Li, S. and Lin, H. (2016). Characterization of triploid hybrid groupers from interspecies hybridization (Epinephelus coioides♀× Epinephelus lanceolatus♂). Aquaculture Research 47(7): 2195-2204.

Joshi, K. D., & Chandra, S. (2020). Early development and parental care in Botia birdii. Fish Reproduction, 10(1), 63-72.

Joshi, R., & Singh, A. (2015). Diversity and Conservation Status of Fish Fauna in the Ladakh Region of India. Environmental Biology of Fishes, 98(12), 2291-2301.

Kottelat, M. (1990). Indochinese Nemacheilines (Pisces: Cypriniformes), a revision of nemacheiline loaches (Pisces: Cypriniformes) of Thailand, Burma, Laos, Cambodia, and southern Viet Nam. Verlag Dr. Friedrich Pfeil.

Kottelat, M. (2012). Conspectus cobitidum: an inventory of the loaches of the world (Teleostei: Cypriniformes: Cobitoidei). The Raffles Bulletin of Zoology, 26, 1-199

Kottelat, M., & Freyhof, J. (2007). Handbook of European freshwater fishes.

Kullander, S. O., & Fang, F. (2004). Fishes of the Western Himalaya: The Nyla and Kashmir Valley. Ichthyological Research, 51(3), 249-254.

Kullander, S. O., Fang, F., & Bannikova, A. A. (1999). Systematics and biogeography of the genus Triplophysa (Cobitoidea, Balitoridae). Ichthyological Exploration of Freshwaters, 10(1), 45-56.

Kumar, P., & Tandon, K. K. (2016). Courtship behavior in Schistura punjabensis: Implications for mate selection. Aquatic Ecology, 50(2), 245-253.

Kumar, S., & Tandon, K. K. (2014). "Environmental Influences on the Growth Rate of Schistura punjabensis in the Himalayan Foothills." Environmental Biology of Fishes, 97(3), 289-298.

Liu, J., Jiang, Y., & Chen, X. Y. (2017). Triplophysa qilianensis, a new species of stone loach (Teleostei: Nemacheilidae) from the upper reaches of the Huangshui River, with remarks on the taxonomic status of Triplophysa stenura (Herzenstein). Zootaxa, 4320(1), 119-130.

Menon, A. G. K. (1987). The Fauna of India and the Adjacent Countries: Pisces. Vol. IV. Teleostei - Cobitoidea. Part 1. Homalopteridae. Zoological Survey of India, Calcutta. pp. 1-10.

Menon, A. G. K. (1999). Check list - Fresh water fishes of India. Records of the Zoological Survey of India, Miscellaneous Publication Occasional Paper No. 175.

Mir, F. A., Yousuf, A. R., & Ahmad, R. (2013). Growth and Population Structure of Triplophysa kashmirensis (Cypriniformes: Nemacheilidae) in a Tributary of River Jhelum, Kashmir. Journal of Ichthyology, 53(3), 221-229.

Nelson, J. S. (2006). Fishes of the World. John Wiley & Sons

Nelson, J.S.; Grande, T.C.; Wilson, M.V. Fishes of the World; John Wiley & Sons: Hoboken, NJ, USA, 2016

Prokofiev, A. M. (2010). Species diversity and distribution of the genus Triplophysa Rendahl, 1933 (Cobitoidea: Balitoridae: Nemacheilinae). Journal of Ichthyology, 50(10), 858-873.

Raina, H. S., & Jamila, M. (1996). Reproductive biology of the loach Triplophysa kashmirensis from a high altitude lake. Hydrobiologia, 337(1-3), 65-70.

Raina, R. K., Pandit, A. K., & Balkhi, M. H. (2015). Influence of Environmental Factors on the Growth Rate of Triplophysa kashmirensis in Kashmir Valley. Hydrobiologia, 757(1), 95-108.

Rendahl, H. (1933). Studien Uber innerasiatische Fische.

Ren, Z., He, D., & Chen, Y. (2012). Fish diversity and conservation in the upper Yangtze River Basin. Acta Ecologica Sinica, 32(2), 441-449.

Rishi, K. K., & Jain, A. K. (2012). "Growth Dynamics of Schistura punjabensis in the Streams of Northern India." Journal of Fish Biology, 81(2), 523-531.

Roy, R., & Dey, S. (2015). Impact of Temperature on the Growth and Development of Botia birdii. Environmental Biology of Fishes, 98(4), 965-973.

Sarkar, U. K., Khan, G. E., Dey, P., & Das, P. (2017). Growth and Population Dynamics of Botia birdii in the Rivers of North-East India. Journal of Fish Biology, 90(5), 1234-1245.

Sharma, R. K., & Joshi, C. B. (2015). Dietary Preferences and Growth Patterns of Schistura punjabensis in Mountain Streams. Indian Journal of Fisheries, 62(1), 92-99.

Shrestha, J., & Subba, B. R. (2017). Reproductive biology and spawning behavior of Schistura punjabensis in Nepal. Journal of Fish Biology, 90(4), 1105-1115.

Siddiqui, M. A. (2018). Courtship and reproductive strategies of Botia birdii. Aquatic Ecology, 52(4), 729-737.

Singh, H., & Singh, V. (2020). Fertilization and early development in Schistura punjabensis. International Journal of Ichthyology, 15(1), 55-62.

Singh, V. K., & Saini, V. P. (2016). Adhesive egg deposition in Botia birdii and its adaptive significance. International Journal of Ichthyology, 12(1), 45-52.

Talwar, P. K., & Jhingran, A. G. (1991). Inland fishes of India and adjacent countries. Oxford & IBH Publishing Co. Pvt. Ltd.

Wang, J., & Li, W. (2013). Reproductive Biology and Longevity of Triplophysa marmorata in Freshwater Streams. Journal of Freshwater Ecology, 28(2), 289-297.

Wang, J., Xie, C., & Li, Z. (2001). Reproductive biology of Triplophysa marmorata in the upper reaches of the Yellow River. Journal of Fish Biology, 59(3), 730-742.

Wu, Y. F. and Wu, C. Z. (1991). The fishes of Qinghai-Xizang Plateau. Sichuan Publishing House of Science and Technology, Chengdu: 25-264.

Wu, T., Xiao, W., & Chen, X. (2018). New insights into the adaptive radiation of Triplophysa (Nemacheilidae) in the Qinghai-Tibet Plateau. Journal of Fish Biology, 92(1), 154-167.

Xie, P., Li, Y., & Xiong, F. (2016). Habitat Preferences and Growth Characteristics of Triplophysa marmorata in the Tibetan Plateau. Hydrobiologia, 765(1), 175-185.

Yang, J., & Cao, W. (2001). Feeding ecology of Triplophysa fish in Yarlung Zangbo River, Tibet. Chinese Journal of Ecology, 20(3), 1-4.

Yang, J. X. (1990). A study on the classification and distribution of the fishes of the genus Triplophysa Rendahl (Cypriniformes: Cobitidae) from China. Acta Hydrobiologica Sinica, 14(2), 97-108

Yang, J., He, S., & Chen, X. (2016). New insights into the phylogeny and biogeography of Triplophysa (Cypriniformes: Nemacheilidae) based on nuclear and mitochondrial DNA data. Molecular Phylogenetics and Evolution, 94, 56-65.

Yang, J., Wu, T. J., Yang, J. X. (2012). A new cave-dwelling loach, Triplophysa macrocephala (Teleostei: Cypriniformes: Balitoridae), from Guangxi, China. Environmental Biology of Fishes 93(2): 169–175.

Zhao, H., Wang, W., & Zhang, H. (2015). Growth and Population Structure of Triplophysa marmorata in the Upper Yangtze River. Journal of Fish Biology, 86(3), 1045-1058.

Zhu, S.Q. (1989). The Loaches of the Subfamily Nemacheilinae in China (Pisces: Cypriniformes: Cobitidae). Nanjing: Jiangsu Science and Technology Publishing House.

Zutshi, D. P., & Gopal, B. (2000). State of Biodiversity in Lakes and Wetlands of Kashmir valley. Environment, Biodiversity and Conservation, Khan, MA and S. Farooq (Eds.). APH Publishing Corporation, New Delhi, India, ISBN-13: 9788176481649, 51-67.

9

Impact of Climate Change on Aquatic Ecosystems and Fisheries in Kashmir Valley Challenges and Adaptation Strategies

Shivbhajan* and Bahni Dhar

Department of Fish Processing Technology and Engineering, College of Fisheries Central Agricultural University (Imphal), Lembucherra, Tripura

Abstruct

The Kashmir Valley, renowned for its stunning landscapes and vital aquatic ecosystems, faces significant threats from climate change. This chapter explores the multifaceted impacts of climate change on the valley's lakes, rivers, wetlands, and fisheries. Rising temperatures and altered precipitation patterns, characterized by decreased snowfall and increased rainfall intensity, are transforming the region's climate. Accelerated glacial melt, resulting in reduced water availability and disrupted river flows, further exacerbates the situation. These changes have profound effects on aquatic ecosystems, including altered water temperatures and quality, habitat loss, and biodiversity decline. Fisheries, particularly those dependent on trout species, are experiencing shifts in population dynamics and productivity due to changing environmental conditions. Addressing these challenges necessitates overcoming data and monitoring gaps, socioeconomic and political constraints, and implementing comprehensive adaptation strategies. Enhancing data collection, promoting sustainable fishery practices, restoring wetlands, and developing climate-resilient infrastructure and policies are crucial steps to mitigate adverse effects and ensure the long-term sustainability of the Kashmir Valley's aquatic ecosystems and fisheries.

Keywords: *Climate change, Aquatic ecosystems, Glacial melt, Biodiversity decline, Sustainable fishery practices*

Introduction

Climate change refers to alterations in the earth's global or regional climates over time, encompassing shifts in atmospheric variability or average conditions over periods ranging from decades to millions of years. The United Nations Framework Convention on Climate Change (UNFCCC) differentiates between "climate change" for human-induced alterations and "climate variability" for other changes. Over the past century, ending in 2005, the average global air temperature near the earth's surface has risen at an estimated rate of 0.74 ± 0.18°C (1.33 ± 0.32°F) (IPCC, 2007). In contemporary discussions, especially within environmental policy, "climate change" often refers to recent climate alterations. Both natural processes and human activities influence the earth's temperature and subsequent climate change. Significant increases in global anthropogenic greenhouse gas (GHG) emissions over the years have been major contributors to global warming. The sun, the earth's primary energy source, has a nearly constant output, but minor variations over extended periods can lead to climate change. Natural processes such as orbital changes (orbital forcing), volcanic eruptions, and greenhouse gas concentrations in the atmosphere also affect the earth's climate. Changes in atmospheric concentrations of greenhouse gases and aerosols, land cover, and solar radiation modify the climate system's energy balance, leading to either warming or cooling.

Volcanic eruptions release several gases, notably sulfur dioxide (SO_2), which forms sulfate aerosols (SO_4) in the atmosphere. Although linking specific weather events to global warming is challenging, increased global temperatures can result in broader changes like glacial retreat, Arctic shrinkage, and rising sea levels worldwide. Variations in precipitation patterns could lead to floods and droughts. Other impacts may include shifts in agricultural yields, the creation of new trade routes, reduced summer stream flows, species extinctions, and the spread of disease vectors. Most global climate change models predict a decline in snowpack on many western mountain ranges, adversely affecting fish populations, hydropower, water recreation, and water availability for agricultural, industrial, and residential use. Partial loss of polar ice sheets could lead to significant sea level rise, altering coastlines and inundating low-lying areas, with the most significant effects on river deltas and low-lying islands. These changes are projected over millennial timescales, but rapid sea level rise within centuries cannot be ruled out. Current climate change models predict a rise in sea surface temperatures between 2 °C and 5 °C by 2100 (IPCC Third Assessment Report, 2001; Done *et al.*, 2003). According to the climate profile document of Jammu and Kashmir, the region has seen an increase in average temperature, with the Kashmir valley rising by 1.45°C and the Jammu region by 2.34 °C over the past two decades. The maximum temperature has increased by 0.5 °C per year in the Kashmir valley and by 0.8 °C per year in Jammu.

The Kashmir Valley, nestled in the Himalayas, is renowned for its breathtaking landscapes, rich biodiversity, and vital aquatic ecosystems. These ecosystems, encompassing lakes, rivers, and wetlands, play a crucial role in the region's ecology and economy. However, the spectre of climate change looms large over this paradise, threatening its aquatic environments and the fisheries that depend on them. This chapter delves into the multifaceted impacts of climate change on the Kashmir Valley's aquatic ecosystems and fisheries, explores the challenges faced by these systems, and proposes adaptation strategies to mitigate adverse effects (Krishan, 2016). Glaciers, acting as natural reservoirs, play a crucial role in maintaining the valley's water balance, particularly during the dry summer months. However, the accelerated retreat of these glaciers due to rising global temperatures is disrupting this balance, leading to reduced water availability and altered flow regimes (Bhat *et al.*, 2019). These changes are further compounded by shifts in precipitation patterns, with an increase in rainfall intensity and a decrease in snowfall, exacerbating the challenges faced by these aquatic systems (Dar *et al.*, 2018). The impact of climate change on aquatic ecosystems in the Kashmir Valley is multifaceted. Rising temperatures are leading to warmer water bodies, which can have cascading effects on aquatic life. Elevated water temperatures can alter the metabolic rates of aquatic organisms, disrupt reproductive cycles, and increase vulnerability to diseases (Shah *et al.*, 2019). Moreover, the proliferation of harmful algal blooms, fueled by warmer temperatures and nutrient runoff, poses a significant threat to water quality and aquatic health (Zargar *et al.*, 2017). Hydrological changes resulting from glacial melt and altered precipitation patterns are leading to habitat loss and fragmentation, particularly in wetlands that serve as crucial breeding grounds for various species. Pandit *et al.* (2021) highlighted a significant reduction in wetland areas in the Kashmir Valley, leading to decreased biodiversity. Species that are highly specialized or have limited ranges are especially vulnerable to these habitat changes.

Fisheries in the Kashmir Valley, primarily centered around species such as the brown trout (Salmo trutta) and snow trout (*Schizothorax* spp.), are deeply affected by climate change. The alteration in water temperatures and flow regimes is shifting the distribution of fish populations, often pushing them to higher altitudes in search of suitable habitats (Nazir and Yousuf, 2018). This vertical migration can disrupt traditional fishing practices and reduce fishery productivity. Changes in water availability and quality also impact the reproductive cycles of fish, leading to lower recruitment rates and declining fish populations. For communities that rely on fisheries for their livelihoods, these changes pose significant economic challenges, threatening food security and economic stability in the region (Rashid *et al.*, 2020). Addressing the impacts of climate change on the Kashmir Valley's aquatic ecosystems and fisheries is fraught with challenges. There are significant gaps in data and monitoring systems, making it difficult to fully understand the extent of climatic impacts and to develop effective management strategies (Mir *et al.*, 2020). Additionally, socioeconomic and political constraints, including limited economic diversification and ongoing political instability, further complicate adaptation efforts (Ahmad *et al.*, 2019). Current climate change models forecast an increase in sea surface temperatures between 2 °C and 5 °C by 2100 (IPCC Third Assessment Report, 2001; Done *et al.*, 2003). According to the climate profile document of Jammu and Kashmir, there has been an overall increase in average temperature, with the Kashmir valley experiencing a rise of 1.45 °C and the Jammu region a rise of 2.34 °C over the past two decades. Additionally, the maximum temperature in the Kashmir valley has risen by 0.5 °C per year, while in Jammu, it has increased by 0.8 °C per year. Figure 1 depicts climate change effects on aquatic ecosystems.

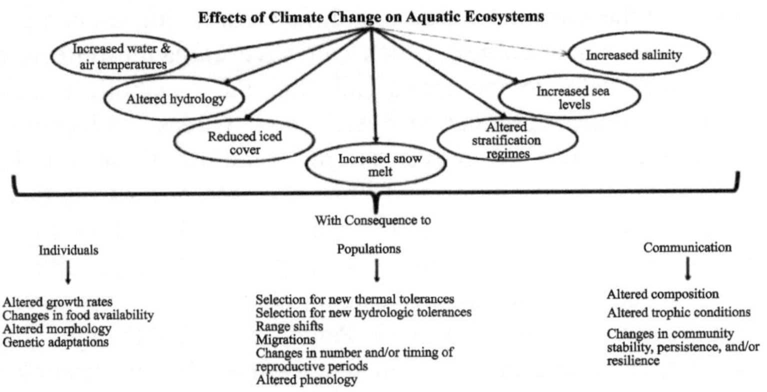

Figure 1. Illustration of climate change effects on aquatic ecosystems and the subsequent individual-population- and community-level responses to these changes. Effective indicators are those that respond to the consequences of climate change.
(*Source:* Hamilton *et al.*, 2010)

Climate Change and Its Manifestations in Kashmir Valley

Historical Climate Patterns

The climate of the Kashmir Valley is profoundly influenced by its mountainous geography, resulting in distinct seasons and substantial climatic variability. Historically, the region has relied heavily on snowfall and glacial melt for its water resources, which feed into its rivers and lakes (Mir *et al.*, 2020). Major water bodies, such as Dal Lake, Wular Lake, and the Jhelum River, are integral to the region's ecology and economy

Recent Climatic Shifts

Over recent decades, the Kashmir Valley has experienced significant climatic shifts, characterized by rising temperatures, altered precipitation patterns, and accelerating glacial retreat. These changes have profound implications for the valley's hydrological dynamics, water quality, and the biodiversity these ecosystems support. Glaciers, acting as natural reservoirs, play a crucial role in maintaining the valley's water balance, particularly during the dry summer months. However, the accelerated retreat of these glaciers due to rising global temperatures is disrupting this balance, leading to reduced water availability and altered flow regimes (Bhat *et al.*, 2019). These changes are further compounded by shifts in precipitation patterns, with an increase in rainfall intensity and a decrease in snowfall, exacerbating the challenges faced by these aquatic systems (Dar *et al.*, 2018).

Temperature and Precipitation Changes

Climate change in the Kashmir Valley is characterized by rising temperatures and altered precipitation patterns. According to a study by Mir *et al.* (2020), the region has experienced a significant increase in average temperatures over the past century, with projections indicating a further rise of 1.5 to 2.5 °C by the mid-21st century. These changes are accompanied by shifts in precipitation, with a notable decrease in snowfall and an increase in the frequency and intensity of rain events (Dar *et al.*, 2018).

Glacial Retreat and Hydrological Impacts

The Kashmir Valley's hydrology is heavily influenced by its glaciers, which serve as vital water reservoirs. However, accelerated glacial melting due to rising temperatures poses a severe threat. Research by Bhat *et al.* (2019) indicates that glaciers in the region have receded by approximately 20% over the past five decades. This retreat not only reduces water availability but also disrupts the timing and volume of river flows, with significant implications for aquatic ecosystems.

Impact on Aquatic Ecosystems

Altered Water Temperature and Quality

Increased water temperatures, resulting from both atmospheric warming and reduced glacial input, have profound effects on aquatic ecosystems. Higher temperatures can alter the metabolic rates of aquatic organisms, affect their reproductive cycles, and increase the susceptibility of fish to diseases (Shah *et al.*, 2019). Additionally, the quality of water bodies deteriorates as warmer temperatures enhance the proliferation of harmful algal blooms and decrease dissolved oxygen levels (Zargar *et al.*, 2017).

Habitat Loss and Biodiversity Decline

Climate change-induced hydrological alterations lead to habitat loss and fragmentation. Wetlands, crucial breeding grounds for many aquatic species, are particularly vulnerable. According to Pandit *et al.* (2021), the Kashmir Valley has witnessed a significant decline in wetland areas, exacerbated by changes in precipitation and increased human activities. This habitat loss threatens the survival of endemic species and reduces overall biodiversity.

Impact on Fisheries

Fish has been a crucial component of the human diet across nearly all nations. It is exceptionally nutritious, offering essential nutrients often lacking in the starchy staples that dominate the diets of many impoverished populations (FAO, 2005; FAO, 2007). Fish accounts for approximately 20% of animal protein consumption globally (Thorpe *et al.*, 2006) and is one of the most affordable and accessible sources of animal protein. In wealthier regions, fish is valued for its oils, which are rich in polyunsaturated fatty acids (PUFAs). Meanwhile, for those at the lower end of the nutritional spectrum, fish is a critical health food due to its proteins, oils, vitamins, minerals, and the benefits derived from consuming small indigenous species (Mohanty *et al.*, 2010a).

In general, fish are unable to maintain a consistent body temperature like mammals. Their body temperature is the same as the water they inhabit. Fish can survive in both cold and hot water, but each species has a preferred temperature range. When fish are exposed to water that is too cold, their metabolism slows down and they become lethargic. Conversely, when the water warms up, their metabolism speeds up, leading to faster growth and reproduction. However, fish with higher metabolism require more food and oxygen. While warmer fish mature faster, they often end up smaller in size. Research shows that aquatic animals raised in warm water are typically smaller than those raised in cooler temperatures. Additionally, warmer temperatures can lead to early hatching and sexual maturity, resulting in smaller sizes compared to fish that grow in cooler waters (Pottinger, 1998). Furthermore, as temperatures rise, some fish may have fewer offspring or become permanently sterile. (Beitinger *et al.*, 2000)

Fish species of Jammu & Kashmir

There are about 138 fish species in state, 93 fish species in Jammu province and 45 in Kashmir province. Many species are in both provinces like *Schizothorax richordsonii, Tor putitora, Glyptothorax, Nemachielus* etc. Commercially important fish species and temperature range presented in (Table 1).

Jammu Province (Commercially important species)	**Temperature range**
Tor Putitora *Tor khudree*	17.5-30.2
Schizothorax richordsonii	18-34.5
Schizothorax longipinnis	------
Schizothorax progastus	-------
Wallago attu	15-35
Bagarius bagarius	22-34
Puntius sarhana	--------
Puntius sophore	--------
Puntius ticto	--------
Labeo dero	23.5-35
Labeo bata	--------
Channa marulius	19-34.4
Channa orientalis	-------

(*Source:* Krishan, 2016)

Fish Population Dynamics

The fisheries of Kashmir Valley, primarily centered around species such as the trout (*Salmo trutta*) and snow trout (*Schizothorax* spp.), are highly sensitive to climate variability. Rising water temperatures can shift the distribution of these species to higher altitudes, leading to changes in population dynamics. A study by Nazir and Yousuf (2018) found that the optimal habitat range for trout has shifted upward by approximately 300 meters over the past three decades, affecting local fishery practices.

Changes in Fishery Productivity

Changes in water temperature, flow regimes, and habitat availability directly influence fishery productivity. Reduced glacial melt and altered river flows can disrupt spawning cycles and reduce the availability of critical nutrients. This impacts fish growth rates and yields, posing significant challenges to the livelihoods of communities dependent on fisheries (Rashid *et al.*, 2020).

Reproductive Cycles and Population Decline

Changes in water availability and quality also impact the reproductive cycles of fish, leading to lower recruitment rates and declining fish populations. For communities that rely on fisheries for their livelihoods, these changes pose significant economic challenges, threatening food security and economic stability in the region (Rashid *et al.*, 2020).

Challenges in Addressing Climate Change Impacts

Data and Monitoring Gaps

One of the primary challenges in addressing the impacts of climate change on aquatic ecosystems and fisheries in the Kashmir Valley is the lack of comprehensive data and monitoring systems. Reliable data on temperature trends, glacial dynamics, hydrological changes, and ecological responses are essential for effective management and adaptation strategies. However, the region's rugged terrain and geopolitical complexities hinder extensive scientific research and monitoring efforts (Mir *et al.*, 2020).

Socioeconomic and Political Constraints

Socioeconomic and political factors further complicate climate adaptation efforts. The Kashmir Valley's economy is heavily dependent on agriculture and tourism, both of which are vulnerable to climate variability. Additionally, political instability and conflicts in the region divert attention and resources away from environmental conservation and climate adaptation initiatives (Ahmad *et al.*, 2019).

Adaptation Strategies

Enhancing Data Collection and Monitoring

Improving data collection and monitoring systems is crucial for informed decision-making. Establishing comprehensive networks of weather stations, glacial monitoring stations, and hydrological sensors can provide valuable insights into climate trends and their impacts. Collaborative efforts involving local communities, government agencies, and research institutions can enhance data accuracy and availability (Mir *et al.*, 2020).

Promoting Sustainable Fishery Practices

Adaptation strategies should prioritize sustainable fishery practices to ensure the resilience of aquatic ecosystems and fish populations. Implementing community-based fishery management approaches, promoting aquaculture diversification, and enforcing fishing regulations can help mitigate the impacts of climate change. Training and capacity-building programs for local fishermen can also enhance their adaptive capacity (Rashid *et al.*, 2020).

Restoring and Conserving Wetlands

Restoration and conservation efforts are essential for maintaining the ecological integrity of aquatic systems. Protecting existing wetlands, rehabilitating degraded ones, and implementing watershed management practices can enhance water retention, improve habitat quality, and support biodiversity.

Engaging local communities in wetland conservation initiatives can foster a sense of ownership and stewardship (Pandit *et al.*, 2021).

Community Engagement and Education

Engaging local communities in conservation and adaptation initiatives fosters a sense of ownership and stewardship, essential for sustainable resource management. Community education programs can raise awareness about the impacts of climate change and promote sustainable practices. Involving communities in decision-making processes ensures that adaptation strategies are culturally appropriate and widely accepted.

Research and Innovation

Ongoing research and innovation are necessary to develop new adaptation strategies and improve existing ones. Research should focus on understanding the specific impacts of climate change on the Kashmir Valley's ecosystems and exploring innovative solutions for adaptation. Collaboration between scientists, policymakers, and local communities can drive the development of effective strategies (Nazir and Yousuf, 2018).

Climate-Resilient Infrastructure and Policies

Developing climate-resilient infrastructure and policies is critical for long-term adaptation. Infrastructure projects, such as constructing climate-resilient water storage and irrigation systems, can help manage water resources more effectively. Policymakers should integrate climate considerations into regional development plans and enforce regulations that protect vulnerable ecosystems from overexploitation and degradation (Ahmad *et al.*, 2019).

Conclusion

The impact of climate change on aquatic ecosystems and fisheries in the Kashmir Valley presents a complex and multifaceted challenge. Rising temperatures, altered precipitation patterns, glacial retreat, and hydrological changes threaten the delicate balance of these ecosystems. The resulting impacts on water quality, habitat availability, and fish populations pose significant challenges to local communities reliant on fisheries for their livelihoods. Addressing these challenges requires a multi-pronged approach that includes enhancing data collection and monitoring, promoting sustainable fishery practices, restoring and conserving wetlands, and developing climate-resilient infrastructure and policies. Through concerted efforts and collaborative strategies, it is possible to mitigate the adverse effects of climate change and ensure the long-term sustainability of the Kashmir Valley's aquatic ecosystems and fisheries.

References

Ahmad, F., Bhat, M. A., & Qayoom, U. (2019). Climate change impacts and adaptation strategies in Kashmir Valley: A review. Journal of Environmental Management, 234, 431-444.

Bhat, M. S., Latief, M., & Shah, A. R. (2019). Glacier retreat in the Kashmir Valley, India: A spatio-temporal analysis. Geosciences, 9(4), 162.

Dar, R. A., Bhat, G. M., & Shah, A. A. (2018). Impact of climate change on water resources of the Kashmir Valley, J&K, India. Journal of Climatology & Weather Forecasting, 6(2), 225.

Done, T., Whetton, P., Jones, R., Berkelmans, R., Lough, J., Skirving, W., & Wooldridge, S. (2003). Global climate change and coral bleaching on the Great Barrier Reef. *Final Report to the State of Queensland Greenhouse Taskforce through the Department of Natural Resources and Mines*, *49*.

FAO (2005). Increasing the contribution of small-scale fisheries to poverty alleviation and food security.

FAO (2007). The state of world fisheries and aquaculture – (2006), 162 p., FAO, Rome.

Hamilton, A. T., Barbour, M. T., & Bierwagen, B. G. (2010). Implications of global change for the maintenance of water quality and ecological integrity in the context of current water laws and environmental policies. *Global Change and River Ecosystems—Implications for Structure, Function and Ecosystem Services*, 263-316.

IPCC (2007). Fourth Assessment Report - Climate Change (2007): Synthesis Report, (2007). IPCC (2001) Climate Change. IPCC Third Assessment Report.

Krishan, R. (2016). Impact of climate change on fish fauna of Jammu and Kashmir. *International Journal of Innovative Research in Science and Engineering*, *2*, 227-236.

Mir, R. A., Rashid, I., & Pandit, A. K. (2020). Climate change and its impact on aquatic biodiversity in Kashmir Himalayas. Water Resources Research, 56(3), e2020WR027083.

Mohanty, S., & Mohanty, B. P. (2009) Global climate change: a cause of concern, NatlAcad

Nazir, R., & Yousuf, A. R. (2018). Effects of climate change on fish habitat and fish health in the Jhelum River, Kashmir Himalayas. Environmental Monitoring and Assessment, 190(4), 212.

Pandit, A. K., Bhat, F. A., & Rashid, I. (2021). Wetland degradation in Kashmir Valley: Consequences and restoration strategies. Wetlands Ecology and Management, 29(1), 15-33.

Parrey, H. A., Shah, S. A., & Bhat, M. S. Analysing Climate Change Manifestations Through People's Perception In Kashmir Valley, India.

Rashid, I., Qadri, H., & Pandit, A. K. (2020). Adaptation strategies for sustainable fisheries in climate-impacted Kashmir Valley. Fisheries Management and Ecology, 27(3), 275-286.

Shah, H., Dar, S. A., & Rather, M. I. (2019). Impact of temperature rise on aquatic ecosystems and fish health in the Kashmir Valley. Journal of Thermal Biology, 82, 59-65.

Thorpe, A., Reid, C., Anrooy, R. V., Brugere, C., & Becker, D. (2006) Poverty reduction Strategy papers and the fisheries sector: an opportunity forgone? J Intl.Dev. 18, 4, 487-517.

Zargar, S., Ahmad, I., & Farooq, S. (2017). Algal blooms and water quality degradation in the Kashmir Valley: Emerging challenges and management strategies. Ecotoxicology and Environmental Safety, 144, 493-500.

10

Traditional Fishing Gear and Crafts Used in Kashmir Valley

Vivek Chauhan[1], Muneshwar Pavan Kalyan[2*], Arya Singh[3] Shyni K.[1] and Jham Lal[2]

[1]*Kerala University of Fisheries and Ocean Studies, Kochi, Kerala*
[2]*College of Fisheries, CAU, Lembucherra, Agartala, Tripura*
[3]*ICAR – Central Institute of Fisheries Education, Mumbai, Maharashtra*

Abstract

The Kashmir Valley's fishing industry has a lot of promise and may greatly boost the state's gross domestic product. However, the unrealized potential must be thoroughly investigated. Since fishing is limited to gathering and selling exclusively using native methods, culture fisheries are still in their infancy. The use of technology interventions in the field of culture fisheries has not yet been investigated. The identification of indigenous fishing techniques, the improvement of conventional techniques, and the implementation of new technology, when needed, are essential for the growth and advancement of cultural fisheries. The Kashmir Valley, which is surrounded by the towering Himalayas, is well known for having a wealth of waterways, including streams, rivers, and lakes that are home to a variety of fish species. Wular Lake is one of the most important habitats among them and plays a major role in Kashmir's fish output. This region's traditional fishing techniques have developed over generations, revealing a profound awareness of aquatic environments and cultural customs in the area. This chapter examines Kashmir's traditional fishing gear, emphasizing methods such handcrafted nets and traps, elaborately carved hooks and lines, and fishing rods made of bamboo and willow. In addition to showcasing the artistry of regional craftspeople, these homemade instruments represent a sustainable fishing method that preserves the natural harmony. Future possibilities for managing and sustaining these traditional fisheries in the face of shifting environmental and socioeconomic environments are reviewed, along with challenges including conservation initiatives and changing fishing techniques. Kashmir's traditional fishing legacy must be preserved to balance cultural continuity with current conservation imperatives as the region continues to embrace development.

Keywords: *Traditional Fishing, Wular Lake, Gear and Craft, Modernization Fishing techniques*

Overview of fishing methods in Kashmir

The water resources of the region of Jammu and Kashmir are abundant, ranging from large lakes in valleys and high altitudes to ponds, pools, streams, marshes, springs, and rivers. Its approximate size is 222236 square kilometers, extending from latitudes 32° 17′ N to latitudes 36° 58′ N and from longitudes 73° 26′ E to longitudes 80° 30′ E. Its whole water spread area is around 32765.3 hectares, or 2% of its overall area (Anon., 2013). Because of their unique natural characteristics, the vast water resources found in lakes, marshes, streams, and rivers are ideal locations for cold water fishing and, as such, have played a significant role in the growth of this sector. These water bodies support diverse populations of both indigenous and exotic fish species, suitable for cultivation as well as those that thrive without human intervention (Mahanta *et al.*, 2009). The Kashmir Valley's water bodies are home to several native and alien fish species (Shah *et al.*, 2020). The Central Asian fauna, which is dominated by the Schizothorax group, makes up the majority of Kashmir's ichthyofauna (Sunder *et al.*, 1978).

Kashmir is endowed with several lakes, including Manasbal, Dal, Wular, and others, each having a unique hydrological environment (Balkhi *et al.*, 1987). Since every lake in Kashmir Valley is located on the Jhelum River's flood plain, the lakes' origins are either tectonic or fluviatile. The Wular Lake, which is situated in District Bandipora and is referred to as the "fish bowl of Kashmir" since it supplies 60% of the fish produced in the Kashmir valley, is said to be the biggest freshwater lake in India (Malik *et al.*, 2018). It spans over 200 square kilometers. The Wular Lake is bordered by several villages all along its shores, including two sizable cities, Bandipur and Sopore, Gurumayun and Choudhury (2009). Rich biodiversity is supported by Wular Lake and the marshes that surround it. It is a significant fishing resource in the valley, providing food for a sizable population that lives on its edges. The lake is the valley's greatest fishing resource and an essential home for migrating water species, including geese, ducks, shorebirds, and cranes (Syed *et al.*, 2020). An understanding of fishing gear is essential for managing fisheries, Target species, fishing location, climate, and water velocity are only a few of the variables that affect how different fishing methods and equipment are used and regulated. Due to the unique characteristics of these habitats, classic fishing techniques that are useless in plains water bodies with slower current may be used in locations with swiftly flowing water, especially in mountainous regions. Understanding the traditional catching practices and gear of a region is crucial for promoting sustainable and responsible fishing practices. With time, traditional fishing techniques have evolved to fit local physical characteristics, including the desired species and goal size of fish. Overfishing is a worrying global trend in which fish are being taken out of aquatic environments at a pace higher than their natural rate of regeneration. This overexploitation has led to a significant decline in global fish populations, a phenomenon often referred to as fish stock crashing. Unsustainable fishing practices are a major contributing factor to this decline (Amos *et al.*, 2018). Although "sustainable fishing" is not required by law, it is a guiding concept for worldwide fisheries management initiatives. Understanding the selectivity of fishing gear is essential to fisheries management. The makeup of catches and the size of target species taken might vary depending on the type of gear used. Therefore, knowledge of gear selectivity is crucial for making informed decisions about how to regulate fishing activities to achieve sustainability goals (McClanahan and Mangi, 2004). Fishing gear used in sustainable fishing techniques must emphasize capturing adult fish at ages while enabling juveniles and subadults to evade capture. This strategy reduces the amount of non-targeted and protected species caught while maximizing productivity (Boopendranath and Pravin, 2009). The fishing techniques that have survived the test of time are the most effective in a certain location or region. Some of the traditional fishing equipment and techniques used in the Kashmir Valley will be covered in this chapter.

Types of hand-crafted fishing gear

In Kashmir, traditional craftsmanship extends to various types of handcrafted fishing gear. Here are some examples:

1. **Fishing Rods (Bamboo and Willow):** This kind of fishing gear is similar to a hook and line setup, except the line is linked to a bamboo pole. Above the hook, the line has a structure that resembles a float, which aids in detecting when the fish is accepting the bait and is ready to be hooked. Bait is made with a blend of wheat flour, baked rice, and ground garbage fish. Most residents of the lake mainly use the approach as a recreational activity (Malik *et al.*, 2018). These rods are ideal for the fishing conditions in the area and are sometimes handmade with elaborate decorations.

Source: https://www.indiamart.com/proddetail/smoked-bamboo-fishing-rod-2495278 897 .html

2. **Reels:** In Kashmir, traditional fishing reels are frequently made of metal, wood, or a mix of the two. With their straightforward design, these reels work well for regional fishing methods.

Source: https://bradajohnson.net/wp-content/uploads/2018/12/Srinagar13.jpg

3. **Hand-picking:** This traditional fishing technique requires high skill and is typically employed in shallow streams with low water levels. Fishermen adept in this method follow the movement of fish, waiting for them to take cover under stones or boulders. With swift precision, the fishermen insert their hands beneath the stones to catch the hidden fish. This technique demands considerable expertise and is primarily used to catch species such as *Pseudecheneis*, *Glyptothorax*, *Barilius*, and certain *Noemacheilus*. It is worth noting that sometimes children can also be seen employing this method in shallow streams characterized by stones and boulders (Srivastava *et al.*, 2002).
4. **Hammering:** Boulders serve as a means of concealment for highland stream fish. Fish concealed behind the stones sustain injuries when a

hammer is used to strike them with force. Fish that are hurt, even smaller ones, are collected.

5. **Gaad hodd (Fishing tunnels):** This technique of catching fish is used in nearby areas of tiny streams that have mud banks and bunds. In the stream's bund, a tiny tunnel or hole is built, and some feed is stored there. The aperture is sealed off when the fish enter the tunnel to feed, and they are gathered from the fish hole (Malik *et al.*, 2018).
6. **Tooker (Basket):** Fish basket traditionally called as "tooker" is one of the extensively used fish traps used in the fast water streams of the Kashmir valley. It is a uniquely crafted basket with a large mouth that taper towards the end, crafted from soft willow tree twigs. It is installed downstream. Fish that are moving with the current become stuck in the taker.
7. **Scare fishing:** This technique is used in small streams, usually with one or two people hitting the banks and water to frighten the fish. The fish are forced downstream and out of their hiding spots by this activity. A big net called a "Khur" is placed at the channel's mouth to catch the fish. The fish get stuck in the net as they are forced downstream (Srivastava *et al.*, 2002).

Commercial methods

These techniques are widely used by local fishermen for commercial fishing, focusing on areas close to the fish markets. But fishermen may also use these techniques in isolated locations, bringing the fish they catch to nearby marketplaces. While some fish farms could try to overfish, if these techniques are used responsibly, there is no inherent risk to fish populations. Fishermen's livelihoods can be preserved while ensuring the sustainability of these fishing activities through appropriate regulation and management.

8. **Nets:** In Kashmir, handwoven fishing nets are a common sight. Cotton or linen are common natural fabrics used to make these nets. Depending on the kind of fish being pursued and the fishing technique used, they come in different sizes.

Source: https://daijiworld.apsouth1.linodeobjects.com/Linode/images3/allwyn_210722_nets.jpg

9. **Traps:** In the waterways of Kashmir, fishermen utilize fish traps made of wicker or bamboo. These traps are made to draw fish into an enclosure that is difficult for them to escape from once they are inside. Fish can be attracted into portable traps with or without bait by entering them through a tiny aperture. One or two compartments with non-returnable, funnel-shaped valves are typical features of traps. In addition to the funnel and chambers, there is a door for recovering captive animals. The procedure calls for the trap to be weighed down with a small amount of concrete, bricks, or flat stone (sinkers), and a buoy line that is fastened to a corner at the chamber end (Reed *et al.*, 1967; Welcomme, 1985).

Source: https://thanhcongcraft.com/wp-content/uploads/2024/02/How-to-make-a-bamboo-fish-trap-2-1200x856.jpg

10. **Bag net (Kurjaal):** This type of dip net, commonly referred to as a "Kurjaal," has a mouth that is either round or triangular. Each of the three dimensions mouth diameter, pouch depth, and bamboo handle length is one meter. "Sagean" refers to another variety with a deeper bag and a wider diameter. Fish around the water fall are captured using these nets.

Source: https://images.app.goo.gl/gx3nv1DKZVEA3o2B8

11. **Narchoo:** Is another type of wounding gear that has three to five spear heads attached to a sturdy bamboo or deodar pole. Its spear is composed of iron, and its length ranges from 3.5 to 5.0 meters. Fish plummets are metal weights with barbed tips that puncture fish as they fall through the bottom; they are mostly used for catching.

Source: https://images.app.goo.gl/eAHBJxfiFo3F3F378

12. **Gill Nets:** These are the best nets to use for fishing in the lake's deepest depths. The gill nets, known as Thani or Ptaji in the local tongue, have a mesh size that ranges from 45 to 75 mm and measure 15 to 40 m long by 1.5 to 3 m broad. The fish fauna's ability to regenerate itself has been negatively impacted by the usage of these nets. It has been noted that because of the drop in fish catches, the communities themselves have placed limits on the usage of smaller mesh size nets. The State

Government also promoted the use of gill nets. Gill net use is prohibited in some areas by the Department of Fisheries due to a decrease in fish catches in the lake (Malik *et al.*, 2018).

Source: https://cdn8.dissolve.com/p/D1233_5_882/D1233_5_882_1200.jpg

13. **Cast net:** In Kashmir, cast net is locally known as "zaal" is a widely used fishing gear in the Dal, Wular, and Manasbal lakes of the valley, typically made of nylon or cotton thread, with nylon being the more common material choice among fishermen. It has a circular shape resembling an umbrella, with diameters ranging from 1.0 to 2.0 meters. The size of the mesh openings varies between 1.2 to 3.0 centimetres from bar to bar. Iron or lead sinkers, weighing approximately 5.0 kilograms, are attached around the peripheral cord of the net. When thrown, the cast net is aimed to land flat on the water's surface, covering the area where fish are present. The weighted edges of the net facilitate its closure around the fish. Casting this gear requires significant skill. In the Kashmir Lake region, fishermen utilize the cast net throughout the year, except for the months from December to February. Different specifications of the local cast nets may vary slightly depending on the preferences and practices of the fishermen (Yousuf *et al.*, 2023).

Source: https://images.app.goo.gl/3pXCbxeDHuRBvLNE7

14. **Hooks and Lines:** The most basic fishing equipment is this one. A baited hook and line are necessary. The shape, size, point type, wire thickness, and shank type of hooks vary significantly. The fish is said to be in the water where the line is cast, and when it consumes the bait, it is brought in. The catch is frequently excellent quality and alive. The size of the hook and the kind of bait determine the kind of fish that is caught. Selective fishing is made possible by the large array of sizes and types of hooks and lures (bait) that may be utilized. Despite these benefits, line fishing requires a lot of labour. Although hooks are often manufactured in large quantities, Kashmiri fishing lines can be crafted by hand. Frequently, these lines are woven using synthetic or natural fibers that are appropriate for the fishing conditions in the area (Read *et al.*, 1967).

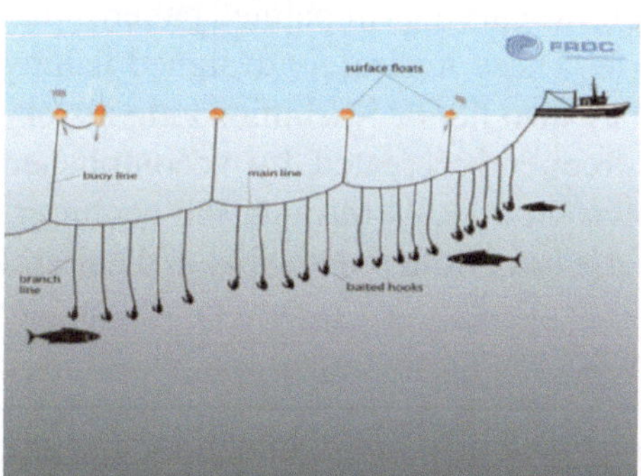

Source: https://fish.gov.au/-/media/fishgov/fishing-methods/hook-andline/drifting_long_line.ashx?la=en&hash=6CE8D29E6A2E65F2EBC2 D88F6621A23F4271950F

15. **Shadow fishing (Tchaai Gaad):** Tchay-e-gard Shikar is a traditional form of fishing in Kashmir. The historical practice is also known as shadow fishing it is specifically associated with Anchar Lake in the saura area of Srinagar (Ganie et al., 2014). The fisherman set out on wooden boats equipped with harpoons they hide under cloth or makeshift umbrellas on the nook of the boat this causes a shadow on the water that attracts fish. The fishermen strike the approaching fish with their harpoons this technique of fishing is usually carried out during winter this is when fish that remain beneath the water surface due to the cold swim towards the shadows for feeding. They use a harpoon, narchoo, and panzer to catch the fish

*Source:*https://www.google.com/url?sa=i&url=https%3A%2F%2Fwww.siasat.com%2Fkashmir-fishermen-in-anchar-lake-brace-up-for-shadow-fishing2966025%2F&psig=AOvVaw2RJYx6J0670bDp6bn4mCXU&ust=1719857788273000&ce=images&cd=-vfe&opi=89978449&ved=0CBQQjRxqFwoTCLiW0Nj3g4cDFQAAAAAdAAAAABAR

16. **Decorative Elements:** Kashmiri craftsmen are skilled at decorating their creations with ornamental aspects, in addition to making useful fishing equipment. This might include elaborate carvings or painted patterns that pay homage to the local way of life. These intricately designed fishing gear are not only functional, but they also honor Kashmir's long history of fine workmanship. They are frequently created by accomplished craftspeople who have passed down their skills through the centuries, guaranteeing that the patterns they use have both practicality and cultural importance.

Destructive Fishing Methods

Local fishermen commonly utilize these methods for commercial fishing, often focusing their activities near fish markets. However, fishermen may also practice these methods in remote areas, transporting the caught fish to markets. While some fish farmers may attempt to overexploit fishing resources, these methods do not inherently pose a threat to fish populations if practiced within limits. Proper regulation and management can help ensure the sustainability of these fishing practices while allowing for the continued livelihoods of fishermen.

18. **Poisonous plant extracts:** Various plant products, including the leaves of Khina (*Madhuka* spp.), Rambans (*Agave americana*), stem bark of Jamun (*Syzygium cumini* L), and oil cake of Cheura (*Sapium* spp.), are utilized to capture fish of various sizes. These plant materials are applied to slow-moving water regions, especially pool habitats, after being finely powdered. The stream water turns murky because of the herbal medicines and mud being thoroughly combined. This method is usually used in the early morning or late at night. The fish in the vicinity behave strangely after these herbal preparations are applied; this could be because of nervous system breakdown or low dissolved oxygen levels. Because of this, the impacted fish become tamed and are simple to gather by fisherman using kadiyali, cast netting, or plain cloth (Singh and Agarwal, 2014).
19. **Dynamite:** This method can be employed throughout the year and in all seasons. While it allows for large-scale fishing in a short period, this method is significantly responsible for the decline in fish abundance and diversity. It poses serious threats to aquatic ecosystems and undermines sustainable fishing practices (Sehga, 1999).
20. **Pesticides:** The most used pesticides are Nuvan, Thiodon, and Malathion. These pesticides are applied in high concentrations to the areas with high probability of catching fishes. The application of pesticides not only damages the ecosystem but also affects the health of human beings (Srivastava *et al.*, 2002).

Techniques for making and using traditional tackle

Kashmiri traditional tackle makers have a strong connection to the natural resources of the area and use meticulous workmanship while creating fishing equipment. For generations, Kashmiri artisans have refined their technique, fusing usefulness, and creativity to produce fishing implements that are appropriate for the region's conditions. This is a thorough examination of the methods used in Kashmir to make and use traditional tackle:

Materials Used

1. **Willow Wood (Tali)**: In Kashmir, willow trees are widely distributed and highly valued for their adaptability and resilience. Tali wood is very popular because it bends without breaking when used to make fishing poles.

Source: https://akm-img-a-in.tosshub.com/indiatoday/images/photo_gallery/202210/1-ap22276410251582.jpg?VersionId=r2J9bZJTC77QaYxOzLr3zBr.Hloo5304&size=686:*

2. **Silk Thread**: Due of their strength and smoothness, silk threads are frequently used in traditional tackle since they don't fray or tangle readily when being used.

Source: https://5.imimg.com/data5/ECOM/Default/2023/2/KL/UX/DD/53296564/1-34-27-ac4c7560-5b99-49d0-aeaa-0389919a83c4.jpg

3. **Metal Hooks**: Usually composed of steel or brass, hooks are designed to resist the demands of fishing in a variety of water conditions.

Source: https://islandfishermanmagazine.com/wp-content/uploads/2020/08/Hooks-Feature.jpg

4. **Natural Baits**: Insects like worms and grasshoppers are good local bait alternatives since they draw fish to Kashmiri waterways.

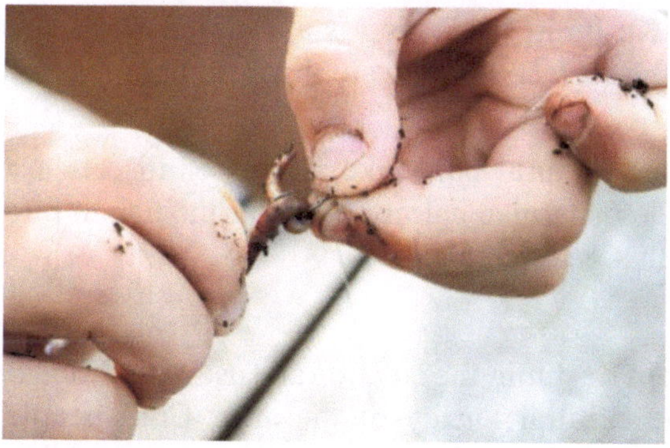

Source: https://thewildprovides.com/wp-content/uploads/2023/01/worm-on-hook-1500-1024x682.jpeg

Techniques for Making Traditional Tackle

1. Fishing Rods (Bumchun)

- **Selection of Willow Wood**: Skilled artisans choose fresh, straight willow branches to make their rods. To decrease moisture content and increase flexibility, the wood is seasoned.
- **Shaping and Carving**: With the use of basic hand tools like chisels and knives, the branches are expertly sculpted. To make sure the rod feels balanced and comfortable in the hand, the artisans pay careful consideration to balance and weight distribution.
- **Binding with Silk Thread**: The rod parts are tightly fastened together with silk thread. This binding offers a decorative aspect with its vibrant colors or motifs in addition to holding the rod together.

2. Fishing Hooks (Kani)

- **Metalworking**: Expert metalworkers use tiny bits of metal to handcraft fishing hooks. They use instruments like pliers and hammers to bend the metal into the required hook shape after heating it to make it pliable.
- **Sharpening**: To guarantee that they enter fish jaws with ease and without inflicting needless harm, hooks are finely honed.
- **Attachment to Line**: To ensure that the hook stays firmly attached during fishing, powerful yet simple-to-tie knots are used to attach hooks to fishing lines.

Using Traditional Tackle

1. Fishing Techniques

- **Fly Fishing (Tambal)**: In Kashmir, fly fishing is a traditional sport in which fish are attracted to the surface of the water by artificial flies that mimic insects.
- **Bait Fishing**: In bait fishing, natural baits like worms or insects are hooked and used as bait. The fisherman calmly waits for the fish to bite before bringing them in with a gentle reel.

2. Cultural Tignificance

- **Artisanal Craftsmanship**: In Kashmir, the creation and use of traditional tackle are ingrained in the cultural legacy of the area. Craftspeople frequently transmit their knowledge to next generations, maintaining methods that have developed over many decades.
- **Environmental Harmony**: Conventional fishing gear is made to cause the least amount of environmental disruption possible, in keeping with a sustainable fishing strategy that protects regional ecosystems.

Challenges Facing Traditional Fishing Practices

Fisheries provide conventional and indigenous peoples not just with food but also with cultural identity and shared unity. The sustainability of the fisheries, including traditional as well as modern fishing methods, is one of the issues associated with fishing activities. The States have the sovereign right to use, protect, and manage marine and fishery resources within their Exclusive Economic Zone (EEZ) in accordance with the 1982 United Nations Convention on the Law of the Sea (LOSC). To prevent overexploitation of the resources and places, several states have put conservation measures in place. Technical restrictions on fishing, such as those related to equipment, time limits, or area restrictions, whether they are included into specific fishing rules and regulations or applied generally, would often not be considered indirect expropriation but rather a control over the use of land Rights (Markowski, 2009). Furthermore, "the cost-effectiveness and social impact of alternative conservation and management measures should be considered in their evaluation." According to CBD (1982), protecting biological variety is a concern shared by all people. Therefore, to preserve the underwater resources, safeguard biodiversity and the environment, and guarantee the recovery of the fishing stock, traditional fisheries must be limited, controlled, and managed much like other fisheries. The traditional use of ecological assets should be safeguarded and promoted in line with customary cultural practices that align with the principles of sustainable use or conservation. Determining what exactly constitutes "traditional" is another difficulty. The term "traditional" has yet to be widely defined; it might refer to many features of people, boats, techniques, technology, fishing grounds, time frames, catches, or fishing activities itself. As a result, the majority of traditional fisherman in Eastern Indonesia continue to fish using "primitive" equipment, and they have recently had to contend with competition from more advanced foreign vessels. Because culture is ever-evolving, traditional fishermen would be permitted to gradually advance their technology, methods, or vessels to prevent technological stability (CBD, 2002). The nature of traditional fishing is flexible and continually shifting.

Future Prospects and Way Forwards

1. **Cultural Preservation and Heritage**: The skilful workmanship used to create fishing nets, hooks, traps, and rods reflects generations-old methods that have been passed down for ages. In Kashmir, these crafts represent communal harmony and cultural identity in addition to being useful for fishing. Initiatives to protect and advance these customs will be needed in the future to make sure they survive modernity and socioeconomic shifts.
2. **Sustainable Fisheries Management**: Sustainable fishing methods naturally correlate with the usage of traditional fishing equipment. Traditional gear reduces environmental effect, in contrast with new methods that could endanger fish populations and ecosystems. To preserve fish populations and aquatic ecosystems in Kashmir's water bodies, future efforts should concentrate on merging traditional knowledge with modern conservation techniques.
3. **Technological Integration**: Although traditional equipment has great cultural and environmental significance, certain technology interventions might be made. Materials innovations can improve durability and efficacy without sacrificing traditional workmanship. Examples of such innovations include eco-friendly fibers for nets and sustainable alternatives for metal hooks. While adhering to current legal restrictions, integrating GPS for navigation or electronic monitoring for capture data might potentially enhance conventional methods.
4. **Economic Opportunities and Tourism**: The traditional fishing gear used in Kashmir has a distinct cultural history that offers prospects for commercial growth. Handmade goods made using age-old methods appeal to a certain consumer base both domestically and abroad. In addition to creating jobs for the local population, efforts to promote traditional crafts through tourist programs like cultural journeys or

artisanal markets may also increase public awareness of the value of maintaining these customs.

5. **Education and Training**: Transferring information and expertise to future generations is necessary to maintain traditional fishing methods. Youth who participate in workshops and educational programs on traditional handicraft will be more equipped to value and carry on these traditions. Working together, craftspeople, environmentalists, and academic institutions may promote creativity without sacrificing cultural authenticity.
6. **Policy and Conservation Initiatives**: To protect traditional fishing rights and encourage sustainable practices, policy frameworks and government backing are essential. To ensure that policies are inclusive and sensitive to cultural differences, conservation projects should include local people in the decision-making process. This strategy boosts local defences against environmental issues while simultaneously protecting traditional fisheries.

Conclusion

A balanced approach that respects cultural heritage, promotes sustainability, grows innovation, and integrates with existing conservation techniques holds great promise for the preservation of traditional fishing gear and crafts in the Kashmir Valley. Kashmir can navigate the challenges of modern development while celebrating its distinctive fishing heritage by adhering to these guidelines.

References

Amos, S. O., and Peter, K. B. (2018). Sustainable artisanal fisheries practices in Nigeria. Oceanography & Fisheries Open Access Journal, 6(1), 8-19.

Anon, 2013. Economic Survey J&K 2012-2013, Directorate of Economics and Statistics, Government of Jammu & Kashmir pp. 652.

Balkhi, M.H., Yousuf, A.R. and Qadri, M.Y. 1987. Hydrobiology of Anchar Lake, Kashmir. Comp. Physiology Ecology 12(3): 131-139.

Boopendranath, M. R. and Pravin, P. 2009. Technologies for responsible fishing-BRDs and TEDs. In Conference: International Symposium on Marine Ecosystems-Challenges and Strategies (9-12).

Convention on Biological Diversity. 2002. Art. 10 (c)

Ganie, P. A., Posti, R., Kunal, G., Bhat, R. A. H. and Sidiq, M. J. (2024). Principle and Applications of Geographic Information System (GIS) in Coldwater Fisheries Development in India. In Aquaculture and Conservation of Inland Coldwater Fishes (pp. 469-495). Singapore: Springer Nature Singapore.

Gurumayum, S. D. and Choudhury, M. (2009). Fishing methods in the rivers of Northeast India.

Mahanta, P. C., Sarma, D., and Joshi, K. D. (2009). Sport Fishing and Livelihood in Upland Himalayan Region. Al India Congress of Zoology

Malik, R., Abubakr, A., and Hussain, N. (2018). Indigenous fish attractants, fishing methods, gears and storage: a study in fishing community of Wular Lake of district Bandipora of Kashmir valley. Journal of Pharmacognosy and Phytochemistry, 7(4), 1751-1755.

Markowski, M. (2009). The international legal standard for sustainable EEZ fisheries management. Towards Sustainable Fisheries Law, 3.

McClanahan, T. R., and Mangi, S. C. (2004). Gear-based management of a tropical artisanal fishery based on species selectivity and capture size. Fisheries Management and Ecology, 11(1), 51-60.

Reed, W., Burchard, J., Hopson, A. J., Jenness, J., and Yaro, I. (1967). Fish and fisheries of Northern Nigeria.

Sehgal, K. L. (1999). Coldwater fish and fisheries in the Indian Himalayas: rivers and streams. Fish and fisheries at higher altitudes: Asia. Food and Agriculture Organization of the United Nations Technical Paper, 385: 41-63.

Shah, M. A., Shah, A. B., and Reshi, Z. A. (2020). Aquatic ecosystems of Jammu and Kashmir State. Biodiversity of the Himalaya: Jammu and Kashmir State, 209-224.

Singh, G., and Agarwal, N. K. (2014). Fishing methods in upper Ganga River system of central Himalaya, India. Journal of Fisheries, 2(3): 195-202.

Srivastava, S. K., Sarkar, U. K., and Patiyal, R. S. (2002). Fishing methods in streams of the Kumaon Himalayan region of India. Asian Fisheries Science, 15(4), 347-356.

Sunder, S., Bhagat, M. J., Joshi, C. B., and Ramakrishna, K. V. (1978). Fishing methods and fish catch composition of Dal Lake, Kashmir (J&K) during 1969–72. J Inland Fish Soc India, 10, 9-18.

Syed, N., Mohite, A., Sadawarte, R., Desai, A., and Shah, T. H. (2020). Design aspects of fishing crafts and gears of Wular Lake of Kashmir, India. Journal of Experimental Zoology India, 23.

Welcomme, R. L. (1985). River Fisheries. Food and Agriculture Organization, Fishery Technical Paper, 262: 1 – 330.

Yousuf, Zarka, Tasaduq Hussain Shah, Farooz Ahmad Bhat, Adnan Abubakr, Bilal Ahmad Bhat, Irfan Ahmad, F. A. Shah, and Nimat Syed. (2023). Fishing gears and craft of Anchar Lake: design specifications and operational aspects. Journal of Experimental Zoology India 26: 2

11

Aquatic Weed Types, Their Distribution Ecological Impact, and Management Strategies in the Water Bodies of Kashmir Valley

Pritam Sarkar[1], Suraj Saha[1], Bhashwati Roy[2] and Puja Rani Basak[1]*

[1]Aquatic Environment and Health Management Division, ICAR – Central Institute of Fisheries Education, Mumbai, Maharashtra

[2] Fish Nutrition, Biochemistry and Physiology Division, ICAR – Central Institute of Fisheries Education, Mumbai, Maharashtra

Abstract

Aquatic weeds, a type of hydrophyte, are integral to aquatic systems and are distributed worldwide, including in India. These weeds thrive in various temperature regions, from tropical to temperate waters. The Kashmir Valley in India is notable for its diverse array of aquatic weeds, particularly those of temperate types, which are found in its lakes, rivers, and wetlands. A total of 191 macrophytes have been reported in the valley, spanning 27 families and 80 genera. This chapter aims to document the types of aquatic weeds found in India, with a special focus on the Kashmir Valley. Additionally, it covers their distribution, ecological roles, negative impacts on ecosystems, and control and management measures.

Keywords: *Aquatic weed, Kashmir Valley, Weed control, Wetlands, Macrophytes*

Introduction

The northernmost territory of the Republic of India, Jammu and Kashmir (latitude 32.28–37.06° and longitude 72.53–80.32°), is situated in the North Temperate Zone, characterized by its temperate-like conditions due to altitudinal variations (220 to 8611 m above mean sea level). The area is rich in biodiversity, with its mountainous terrain creating a beautiful aquatic landscape that is a key feature of its idyllic scenery. The aquatic resources over the valley primarily consist of high-altitude lakes, with wetlands and rivers following. These resources are more prevalent at higher altitudes, with decreasing occurrence at mid and lower altitudes. The freshwater bodies in the valley support a diverse array of macrophytes, phytoplankton, and zooplankton (Shah *et al.*, 2020). Aquatic weeds, or hydrophytes, are found in water bodies worldwide, including those in India. Aquatic weeds are generally described as unwanted plants in the aquatic ecosystem. They occupy various ecological niches within these water bodies and play crucial roles in their ecosystems. India is home to numerous native vascular aquatic plants, as well as various non-native species. These aquatic plants include monocots, dicots, periphyton, and angiosperms. They serve as primary producers in the water bodies, contributing to oxygen production and offering nutritional and medicinal benefits. Despite their importance to the aquatic ecosystem, excessive growth of aquatic weeds can pose significant risks to aquatic organisms, particularly fish diversity in the affected areas.

The aquatic macrophytes of the Kashmir valley have been documented extensively by various authors over time, continuously updating the knowledge of the regional aquatic flora. Based on the recent finding of Ganie *et al.* (2020), the valley currently hosts 191 species of macrophytes, both aquatic and semiaquatic. These plants belong to 27 families and 80 genera, distributed across 22 dicotyledons, 12 monocotyledons, and 3 pteridophytes. Of these, 92 species are fully aquatic, including free-floating and submerged types, while 99 species are semiaquatic and emergent. In terms of diversity, the highest number of macrophyte species (160) has been reported from Kashmir, followed by 94 species in Jammu and 44 species in Ladakh. The distribution of aquatic weeds in the Kashmir valley includes prominent water bodies such as Dal Lake, Wular Lake, and Manasbal Lake. These lakes support a variety of macrophytes, contributing to the rich biodiversity of the valley. Dal Lake, for instance, is known for its extensive growth of *Hydrilla verticillata* and *Potamogeton natans*, while Wular Lake hosts species like *Nymphaea alba* and *Trapa natans*. The variety of aquatic plants in these water bodies is crucial for maintaining the ecological balance and health of the aquatic ecosystems of the valley. The diverse array of aquatic weeds in the waterbodies of the valley acts as a "double-edged sword" for the ecosystem. On one hand, these plants provide oxygen, maintain the nutrient cycle, and serve as food for some phytophagous fishes. On the other hand, their overabundance can create mosquito breeding grounds, absorb vital nutrients from the water, release toxic chemicals, and cause dead zones at night by consuming oxygen for respiration. Therefore, it is crucial to control and manage the population of aquatic weeds to ensure the wellbeing and sustainability of the aquatic ecosystem.

Types of Aquatic Weeds

India as a megadiverse country holds diverse species of aquatic plants. Aquatic weeds are primarily two types such as algae and flowering plants (Macrophytes) (Figure 1). Further macrophytes are subdivided into four groups based on the habitat.

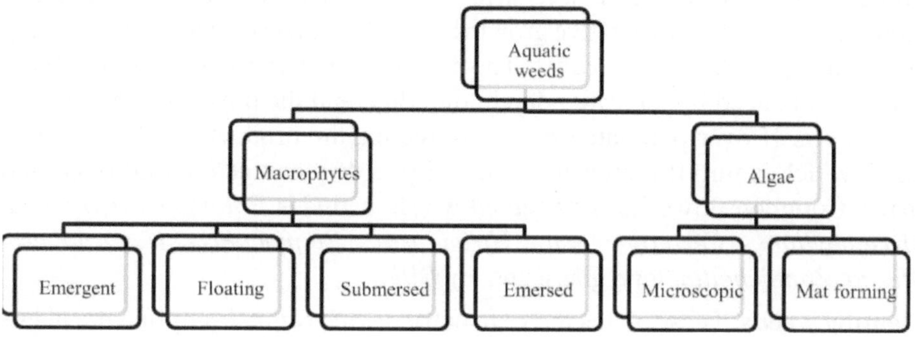

Figure 1: Types of aquatic weed

A. Algae

Microscopic algae, also known as phytoplankton, are single-celled organisms that color the water green, yellow, yellow-green, or brown. A heavy growth of algae can form a "bloom," leading to eutrophication. In still water, blooms can produce surface scum and create green, red, black, or oily streaks. During the day, they convert solar energy into food and remove CO_2 from the water through photosynthesis, producing oxygen as a byproduct. At night or on cloudy days, they release CO_2 into the water through respiration and consume oxygen. In aquatic body often mat forming algal groups are predominant. They are composed of single cells linked end to end, forming nets, branched filaments, single threads, or upright stem-like structures with whorled branches or forked leaves. Typically starting in the spring, they grow into mat-like formations around the bottoms and margins of bodies of water. Significant genera of filamentous algae includes *Pithophora, Hydrodictyon, Spirogyra, Nitella,* and *Chara.*

B. Macrophytes (Hydrophytes)

Unlike microalgae aquatic plants or hydrophytes contain vascular tissues, making them more complex than algae. Vascular structures allow these plants to transport nutrients, water, and other resources throughout their bodies. Macrophytes are widely distributed over the country and common types with representative examples are revealed in (Table 1). Weed macrophytes can be categorized into four major groups based on their habitat.

1. Submerged weeds
2. Floating weeds
3. Emergent weeds or marginal or shoreline weeds
4. Emersed weeds

Submerged Weeds: These weeds have roots anchored at the bottom and produce most of their vegetative growth below the water surface. They typically have true stems, roots, and leaves. The density and quantity of these weeds are mainly influenced by the water depth, turbidity, and the physical properties of the bottom. If over populated they may reduce the production of planktonic algae, which in turn lowers fish productivity as they compete for nutrients and space. Common flowering submerged weeds includes *Potamogeton, Elodea, Myriophyllum, Ceratophyllum, Utricularia, Ranunculus, Heteranthera, Alisma, Zannichellia,* tape grass, and *Hydrilla.*

Floating weeds: These plants thrive in nutrient-rich water. Many aquatic plants have leaves that float on the water surface, either singly or in rosettes. They possess true leaves and roots. Some have leaves that float up to the surface, while others are rooted in the bottom mud. They reproduce very rapidly under favorable condition and are the most troublesome of aquatic weeds Free-floating plants, such as watermeal, duckweed, and mosquito fern, can form mats that completely cover a pond surface, blocking out underwater life and reducing the amount of oxygen in deeper waters. These plants are typically found in still or stagnant waters. Common genera of these coating weeds includes *Brasenia, Nelumbo, Nechamendra, Lemna, Nymphaea, Eichhornia,* and *Pistia.*

Emergent weeds: Emergent weeds, which grow on saturated soil above the water surface, make up the majority of marginal weeds. They extend from wet coastal areas into water 60 to 90 cm deep. This group, which includes herbs, shrubs, trees, and some grasses with broad leaves, is the most extensively distributed group of rooted aquatic plants. These plants primarily consist of broadleaved, grass-like varieties and often spread rapidly through both seeds and subterranean rhizomes. Pond margins host a wide range of weeds, including marshes (*Eleocharis*), sedges (*Cyperus*), grasses (*Paspalidium*), water ipomoea (*Ipomoea aquatica*), and water primrose (*Jussiaea*). Common grass-like plants include reeds, cattails, bulrushes, and spikerushes.

Emersed Weeds: These plants have stems and leaves growing at or extending above the water surface, rooted in the bottom muck. This group includes some weeds with broad leaves measuring 5 to 50 cm in diameter, and others with long, narrow leaves resembling grasses, less than 3 to 15 cm in width; the latter are known as reeds. Their blossoms and foliage found floating at the surface of the water and having depth less than five feet. Some common genera of these emergent weeds include *Myriophyllum, Jussiaea, Nelumbo,* and *Nuphar.*

Table 1: Types of aquatic weeds abundant in India (Kawade *et al.*, 2023)

Type		Common name	Scientific name	Family
Emergent Weeds		Cattail narrowleaved	*Typha angustata*	Typhaceae
		Common reed	*Phragmites communis Trin*	Poaceae
		Pickrel weed	*Pontederia cordata L*	Pontedericeae
		Watergrass	*Commelina benghalensis L*	Commelinaceae
		Umbrella plant	*Cyperus difformis L*	Cyperaceae
		Besharam	*Ipomea carnea Jacq.*	Convolvulaceae
Floating Weeds	Free Floating	Water hyacinth	*Eichhornia crassipes (Mart) Solens*	Pontederiaceae
		Water fern	*Salvinia auriculata (Mitch) Syn*	Salviniaceae
		Water lettuce	*Pistia stratiotes L*	Araceae
		Duck weed	*Lemna minor*	Lemnaceae
		Water velvet	*Azolla imbricata waxai*	Salviniaceae
	Rooted Floating	Arrowhead	*Sagittaria guayanensis HBK*	Alismataceae
		Nilkalmi	*Ipomea hederacea*	Convolvulaceae
		Lotus	*Nelumbo nucifera G*	Nymphaceae
		White water lily	*Nymphaea alba L*	Nymphaceae
		Horned pond weed	*Zannichellia palustris L*	Zannichelliaceae
Submerged Weeds	Shallow water	Blue green algae	*Anabaena spp*	Nostocaceae
		Cottonmat type algae	*Cladophora spp*	Cladophoraceae
		Wet wool type algae	*Pithophora spp.*	Chloraphyaceae
		Slimy green algae	*Spirogyra spp.*	Chloraphyaceae
		Musk grass	*Chara zeylanica*	Characeae
		Stone wort	*Nitella hyalina*	Characeae
		Naiad	*Najas minor All.*	Najadaceae
		Eel weed	*Vallisneria spiralis*	Najadaceae

		Curly leaf pond weed	*Potamogeton crispus*	Potamogetonaceae
	Deep water	Eurasion water milfoil	*Myriophyllum spicatum L*	Holorhagaceae
		Hydrilla	*Hydrilla verticillata Royle*	Hydrocharitaceae
		Elodea	*Elodea canadensis*	Hydrocharitaceae
		Bladderwort	*Utricularia flexuosa Vahl*	Lentibulariaceae
Emersed weeds		Water milfoil	*Myriophyllum spp*	Halorragaceae
		Water Primrose	*Ludwigia adscendens*	Onagraceae
		Indian lotus	*Nelumbo nucifera*	Nelumbonaceae
		Water lily	*Nuphar luteum*	Nymphaceae

Distribution

The distribution of reported species of aquatic weeds in the Kashmir valley represents a fascinating interplay of natural habitats, environmental conditions, and human activities within this geographically diverse region. Located in the northern region of the Indian subcontinent, the valley is famed for its stunning scenery, featuring high-altitude lakes, winding rivers, and verdant marshlands. These aquatic environments support a variety of plant species, some of which have become notable as aquatic weeds due to their ecological impact. The reported aquatic weeds reported from the valley (Table 2) are diverse and holds ecological importance. The distribution of these aquatic weeds is influenced by several factors, including water chemistry, nutrient levels, water depth, and human activities such as agriculture, urbanization, and recreational use of water bodies. For instance, *Lemna minor* and *Spirodela polyrhiza* are commonly found in nutrient-rich lakes and ponds, forming dense mats that can affect water flow and oxygen levels. In contrast, species like *Phragmites australis* and *Typha angustifolia* thrive in marshy areas and along lake edges where water levels fluctuate seasonally. The water bodies serve as critical habitats for a diverse range of aquatic plants, including both native species adapted to the local climate and introduced species that have established themselves over time. Aquatic weeds in this context include floating, submerged, and emergent species, each occupying distinct niches within the aquatic ecosystem. Among the various lakes, some of the most diverse are highlighted, along with the distribution of aquatic weeds across these lakes.

- **Dal Lake**: As one of the most iconic water bodies in the region, Dal Lake (near Srinagar) is renowned for its intricate network of floating gardens (*Radh*) and houseboats. It hosts a variety of aquatic plant species, including notable weeds such as *Lemna minor* and *Salvinia natans*. The presence of these species reflects the nutrient-rich environment and cultural importance of the lake.

- **Wular Lake**: Being the largest freshwater lake in India, Wular Lake (near Bandipora) supports a complex ecosystem that includes diverse aquatic vegetation. Species like *Phragmites australis* and various species of pondweeds (*Potamogeton* sp.) are reported here, thriving in the marshy edges and shallow waters of the lake.
- **Manasbal Lake**: Known for its natural beauty and ecological significance, located in Ganderbal district, Manasbal Lake harbours aquatic plants such as *Nymphaea* sp. and *Typha angustifolia*, which contribute to the biodiversity and scenic appeal of the lake.
- **Anchar Lake**: Close to Srinagar, Anchar Lake also provides habitat for *Polygonum amphibium, Nymphoides peltala, Nitella hyalina*. These can be found in the shallow areas and along the edges where water conditions are favourable.

Table 2: Commonly reported aquatic weeds from the Kashmir valley (*source* – Kaul *et al.*, 1976; Ganie *et al.*, 2020; Munibul, 2022)

Type	Reported Species
Submerged	*Potamogeton lucens*
	Potamogeton pectinatus
	Potamogeton perfoliatus
	Myriophyllum spicatum
	Ceratophyllum demersum
	Utricularia flexuosa
	Nitella hyalina
	Potamogeton natans
	Chara spp.
	Najas spp.
Floating	*Polygonum amphibium*
	Nymphoides peltala
	Trapa natans
	Marsilia quadrifolia
	Nelumbo nucifera
	Salvinia natans

Emerged	*Phragmites australis*
	Typha angustifolia
	Alisma plantigo aquatica
	Scirpus lacustris
	Butomus umbellatus
	Sagittaria sagitifolia
	Eleocharis palustris
	Phalaris arundinacea
	Paspalum dilatalum
	Sparganium electum
Floating	*Spirodela polyrhiza*
	Ricciocarpus natans
	Lemna gibba
	Lemna trisulca
	Lemna minor

Role of Aquatic Weed in Waterbodies

Food Source

Feeding is an integral part in the aquatic ecosystem is crucial for the growth, health, and overall productivity of aquatic organisms.

- **Water spinach (*I. aquatica*):** Water spinach, known locally as 'Kolmi,' is widely abundant in India and other regions across Asia (Nandi *et al.*, 2023). Due to its exceptional nutritional profile, water spinach is highly suitable for inclusion in aquafeed formulations. According to Adedokun *et al.* (2017), water spinach leaves contain approximately 26% crude protein, 6% crude fat, 13% ash, 8% crude fiber, 48% carbohydrates, and 5% moisture by dry weight. Moreover, Yousif *et al.* (2019), found that water spinach can substitute the use of fish manure by 25% in case of Nile tilapia farming.
- **Duckweed (*L. minor*):** Duckweed is highly valued for aquafeed formulation due to its ubiquitous abundance, high proliferation rate, nutrient rich nature and economical values. It boasts a protein content exceeding 40 g 100 g^{-1} on a dry matter basis (Stadtlander *et al.*, 2019), with a fiber content of 5 g 100 g^{-1} in dry weight (Irabor *et al.*, 2022). Additionally, duckweed contains significant quantities of amino acids, phosphorus, and potassium, making it a valuable nutritional resource.
- **Water lettuce (*P. stratiotes*):** Water lettuce, is an aquatic plant that resembles a floating flower on the water surface. The approximate composition of water lettuce leaves includes 7% crude protein, 2% ether

extract, 35% ash, 18% fiber, 38% carbohydrate, and 5% moisture. In contrast, the roots are rich in minerals and contain approximately 3% crude protein, 2% ether extract, 45% ash, 21% fiber, 45% carbohydrate, and 5% moisture (Wasagu *et al.*, 2013).

- **Water hyacinth (*E. crassipes*):** It presents a promising opportunity as a cost-effective source of protein for fish diets. Water hyacinth contains crude protein levels ranging from 10.1% to 11.2%, crude lipid levels from 1.1% to 1.8%, crude fiber levels from 26.1% to 27.4%, and crude ash levels from 12.3% to 12.4% throughout the entire plant (excluding the roots), indicating its high nutritive value (Hossain *et al.*, 2015).
- **Mosquito fern (*A. pinnata*):** Mosquito fern is recognized as a significant protein source, containing more essential amino acids compared to wheat flour, maize, offal, and similar sources. According to Mangesh *et al.* (2018), the proximate composition of mosquito fern is approximately 22.25% crude protein, 2.45% crude fat, 25.50% crude ash, and 11.19% crude fiber on a dry weight basis, alongside substantial levels of minerals, amino acids, and retinol.
- **Water lily (*N. nouchali*):** Water lilies are perennial aquatic plants belonging to the Nymphaeaceae family, distinguished by their ability to float with leaves and blossoms resting on the water surface. They are recognized for their nutritional richness, phytochemicals, phenolic compounds, and carbohydrates. Water lilies also provide significant levels of vitamins and minerals, comprising approximately 76.5% carbohydrates, 10.76% crude protein, 2.4% crude fat, and 3% crude fiber (Anand *et al.*, 2019).

Oxygenation

Aquatic plants also engage in photosynthesis, much like terrestrial plants. This process, which mostly takes place during the day, involves their absorbing carbon dioxide and releasing oxygen. The water absorbs this oxygen, raising the oxygen concentrations. Algal bloom control is aided by aquatic vegetation. Because algae use oxygen at night or during their decomposition, excessive development of algae can result in oxygen shortage. By competing with algae for nutrients, aquatic plants support a healthy ecology. Among the many creatures that aquatic plants support and nurture are fish and invertebrates. They maintain a varied variety of creatures, which promotes the general health and balance of the ecosystem.

Phytoremediation

Aquatic plants naturally absorb various pollutants from the water. Using these plants to eliminate various high molecular weight metals and other contaminants is an effective and economical method. Constructed wetlands incorporating aquatic plants have been widely utilized globally for wastewater treatment (Gorito *et al.*, 2017). Hydrophytes often produce vast root systems, making them highly effective for accumulating contaminants in both their roots and shoots. Certain floating hydrophytes are particularly best-known for the ability to remove metals from polluted environments. These include water hyacinth (*Eichhornia crassipes*), water ferns (*Salvinia minima*), duckweeds (*Lemna minor, Spirodela intermedia*), water lettuce (*Pistia stratiotes*), and water cress (*Nasturtium officinale*). In free-floating aquatic plants, heavy metals are actively transported to different sections of the plant body from the roots. The direct interaction of the plant body with the pollutant medium is linked to passive transport. Heavy metals mainly accumulate in the top sections of plants during passive transport. Water lettuce is adept at reducing BOD, COD, DO, pH, ammonia, nitrite, nitrate, and phosphate, particularly excelling in removing zinc and mercury from industrial wastewater. Duckweed removes a broad spectrum of heavy metals, contaminants, pesticides, and nutrients, inhibiting algae and fungi growth, and effectively reducing nitrogen levels. Submerged aquatic plants absorb heavy metals through their leaves, removing them from both water and sediments. Emergent aquatic plants, living in submerged soils, accumulate heavy metals at varying rates, often storing them in their roots but sometimes dispersing them through aerial parts, detoxifying them at the cellular level.

Consumption and Medicinal Value of Aquatic Weed

Aquatic and semi-aquatic plants in the Kashmir Valley are collected from both wild and cultivated flooded areas, aquaculture setups, and agricultural fields to fulfil various needs. While many plants are used for medicinal purposes, only a limited number are utilized for human consumption, often restricted to local use due to traditional knowledge. Major plant species consumed by humans include Indian lotus (*Nelumbo nucifera*), water caltrop (*Trapa natans*), watercress (*Nasturtium officinale*), water spinach (*Ipomoea aquatica*), wild rice (*Zizania* sp.), and wild taro (*Colocasia esculenta*). Additionally, various species are used for food, utilizing parts such as stems, leaves, roots, rhizomes, flowers, flower heads, and fruits.

- **Indian Lotus (*Nelumbo nucifera*)**

  The Indian lotus is a crucial edible aquatic plant widely cultivated for its medicinal, nutritional, and ornamental value. It is grown extensively in Asia and Australia. The rhizomes are used in soups, deep-fried, or stir-fried dishes. Rich in dietary fibre, vitamins, and nutrients, lotus roots are also made into pickles or added to salads and other foods. Beyond its culinary uses, the Indian lotus exhibits strong antioxidant properties in its rhizomes, stamens, leaves, and seeds, with bioactive compounds found throughout the plant. This plant is extensively consumed in the Kashmir Valley, contributing to local diets and traditional medicine.

- **Taro (*Colocasia esculenta*)**

  It's a perennial, aquatic weed grown in many areas of the globe, including North African, South Asian and Pacific countries. Taro leaves are rich in carotene, potassium, calcium, phosphorous, iron, riboflavin, thiamine, niacin, vitamin A, vitamin C, and dietary fibre. Edible parts of taro include corms, cormels, leaves, stalks, and inflorescence, providing balanced nutrition (Akwee *et al.*, 2015). Taro is also used in traditional medicine for treating arterial hypertension, liver problems, ulcers, snakebites, and rheumatism (Safo-Kantaka, 2004).

- **Wild Rice (*Zizania palustris*)**

  Wild rice, also known as Canadian rice, Indian rice, or water oats, consists of four species: *Zizania palustris*, Z. *aquatica*, Z. *texana*, Z. *latifolia*. It is valued for its high protein content (6.5%), which is higher than its fat and carbohydrate content (compared to brown rice), making it an important cash and subsistence food cereal (Oelke, 1993). It also has higher levels of lysine and methionine compared to other cereals. Wild rice is abundant in gamma-oryzanol, a natural antioxidant that includes ferulic acids, sterols, triterpene alcohols, and vitamin E (Aladedunye *et al.*, 2013). Additionally, it has capacity to replace white rice and processed starch to help prevent chronic metabolic diseases. In the Kashmir valley, wild rice is appreciated for its nutritional benefits and is increasingly incorporated into local dishes, contributing to a diverse and healthy diet.

- **Watercress (*Nasturtium officinale*)**

  Watercress is a perennial aquatic herb that thrives along the shores of rivers, lakes, reservoirs, and winding streams. It is cultivated globally in both temperate and subtropical regions. It is known for its peppery taste akin to many vegetables (Rose *et al.*, 2000), watercress is a nutritional powerhouse, suppling minerals and vitamins and traditionally has been used to combat scurvy. Watercress contains medicinally valuable bioactive compounds and volatile oils, offering benefits such as anti-cancer properties and stimulating oxidative enzymes to combat DNA breakdown (Casanova *et al.*, 2013). In culinary use, it is popular in salads, sandwiches, soups, and vegetable drinks.

- **Water Caltrop (*Trapa natans*)**

  Water caltrop is a significant food crop in Africa, Australia, and various other regions around the world, including Asian countries like China and India. The seeds can be consumed either raw, roasted, or boiled. Boiling the seeds or grinding dried nuts into flour can eliminate toxins and water flukes present in the raw seeds (Aasim *et al.*, 2018). Water caltrop is low in fat, cholesterol-free, and high in fiber, which supports the growth of probiotic bacteria. Additionally, it contains flavonoids, glycosides, saponins, phytosterols, tannins, and iodine.

- **Water Spinach (*Ipomoea aquatica*)**

  Water spinach is a floating edible aquatic plant indigenous to the Indian subcontinent, Southeast Asia, and Eastern Asia. Although the whole plant can be eaten as a vegetable, the leaves are the most commonly consumed part, despite their tendency to deteriorate quickly after cooking. The leaves offer a mild, enjoyable, and subtly sweet taste, and are packed with essential minerals and vitamins. They have a crude protein content of 18–34%, and the young leaves are frequently used in various vegetable dishes (Aasim *et al.*, 2018).

Negative Impact of Aquatic Weed on the Aquatic Environment

The Kashmir valley is a lacustrine basin located in the intermontane depression between the lesser and Greater Himalayas, rich with various freshwater ecosystems. The Kashmir Himalayan region is known for its rich biodiversity due to its unique geological history and eco-climatic regimes (Khan *et al.*, 2004). Aquatic weeds, particularly invasive species, pose significant threats to the freshwater ecosystems of the Kashmir valley. The waterbodies like Dal Lake, Wular Lake, and the Jhelum River, faces specific challenges due to the negative impacts of aquatic plants.

- **Ecological disruption and biodiversity Loss**

  In the typical water bodies of the Kashmir Himalayas, there are 129 alien aquatic species from 68 genera and 42 families (Shah and Reshi, 2014). Eurasian Watermilfoil and Hydrilla, these invasive aquatic plants have established themselves in the water bodies of the Kashmir valley. The largest families of invasive aquatic plants, Cyperaceae and Potamogetonaceae can outcompete and displace native aquatic vegetation in the valley. The rapid spatial spread of invasive aquatic plants often occurs at the expense of traditionally valued native species.

According to Reshi *et al.* (2008), cumulative invasions have transformed diverse and heterogeneous ecosystems into more homogenous systems. They outcompete native plant species, altering the natural ecosystem balance, diminishing the ecological niches available for other organisms, disrupting food webs and ecological processes, and ultimately declining native biodiversity.

- **Water Quality Degradation**

  Excessive growth of aquatic plants, often driven by nutrient runoff from agriculture and urban areas, leads to eutrophication. Fish and other aquatic creatures are harmed by this process, which lowers the oxygen content of the water. Additionally, toxic algal blooms caused by nutrient enrichment can endanger people, animals, and wildlife. Excessive growth of aquatic weeds in rivers and lakes can impede water flow, increasing the risk of flooding during heavy rains in local areas. Stagnant water, due to dense plant growth, can become breeding grounds for mosquitoes, increasing the risk of waterborne diseases like malaria and dengue. Therefore, the overgrowth of aquatic plants can impact the effectiveness of local sanitation systems, leading to water quality issues that affect public health.

- **Loss of Aesthetic and Recreational Value**

  The scenic beauty of the Kashmir valley is largely attributed to its diverse and vibrant freshwater ecosystems. The Kashmir valley is a major tourist destination, and water-based activities like shikara rides, boating, and fishing are popular. However, the aquatic vegetation can impede swimming, boating, and other recreational activities, reducing the appeal of these water bodies to tourists, leading to a decline in tourist satisfaction and potentially reducing revenue from tourism.

- **Impact on Livelihood and Economy**

  The overgrowth of aquatic plants can adversely affect the local fishing industry. Dense vegetation can alter water quality and hinder fish movement and breeding, reducing fish populations and impacting the livelihoods of local fishermen. Moreover, managing invasive aquatic plants can be costly and often requires significant financial resources and labour. Water bodies clogged with aquatic plants can lead to reduced water availability for irrigation, affecting agriculture, which is a significant part of the local economy. Many aquatic plants are naturally invasive because they can quickly expand across the waterbody (Richardson *et al.*, 2000). The invasion of aquatic weeds in the Kashmir valley is multifaceted, impacting ecological, aesthetic, recreational, and economic aspects of the region. The transformation of diverse ecosystems into homogenous ones, the displacement of native species, the loss of scenic beauty, and the reduction in ecotourism potential all contribute to the overall degradation of freshwater habitats. The escalating management costs, coupled with intensified human activities, have significantly compounded the complexity of these issues.

Management and Control of Aquatic Weeds

Aquatic plants play a crucial role in aquatic ecosystems by providing food and shelter to fish, animals, and other aquatic life forms. However, as described in the earlier sections certain invasive aquatic plants can be problematic as they can survive in less water and can disrupt the healthy functioning of native aquatic ecosystems. Weed control is a critical part of ecosystem management, especially for alien aquatic weeds. In Kashmir, significant attention is focused on managing species such as *Eichhornia crassipes, Ipomoea aquatica, Typha angustata, Ceratophyllum demersum, Salvinia molesta, Nelumbo nucifera, Alternanthera philoxeroides, Hydrilla verticillata, Vallisneria spiralalis, Chara* sp.*, Nitelia* sp., and *Potamogeton* sp. due to their wide distributions and often negative impacts. Some effective control and management processes which are followed widely are described below.

- **Manual or Mechanical Control Method**

  Traditional methods like hand-picking, pulling, and scything emergent and marginal weeds are considered suitable for small water bodies. However, one major limitation of these methods is their impracticality and inefficiency in dealing with large infestations. For rooted submerged weeds in shallow areas; mechanical removal can be achieved by:

  - Raking with a log weeder, typically made of wood with iron spikes and barbed wire.
  - Using bamboo poles with a cross attachment at the base or a broom fork with toothed prongs for turning.
  - Employing sickles and long-handled forks to rake or chop vegetation.

In cases of deeply rooted underwater vegetation, a motorized winch is employed for effective removal. A portable mechanical system capable of removing both floating and submerged weeds at a rate of 1-1.5 hectares per day has been developed at the ICAR – Central Institute of Fisheries Technology (ICAR – CIFT) (Kawade *et al.*, 2023).

- **Biological Control**

  Certain types of weeds can be suppressed by using specific fish species that consume plants. The most commonly employed ones are gourami, common carp, tawes, tilapia, and grass carp. Two more species that are used to control aquatic vegetation are ducks and geese. The main benefits are that treating the water does not harm plants that are irrigated with it, and it is easy to apply. The submerged vascular aquatic weeds can be effectively controlled by triploid grass carp. Rather than only regulating certain vegetation in a pond, grass carp are typically employed to manage all of the flora there. When fish are lost, replacement grass carp stocking is required. Most of the species are subdued by tilapia, which are stocked in the spring.

- **Use of insects:** Since aquatic weeds make up 99% of the population, natural biological regulators are ineffective. The water hyacinth is the most common aquatic weed globally. Insect species such as *Agasicles hygrophila* and *Vogtia malloi* are used to control aquatic weeds. This issue is particularly significant in the Indian subcontinent, Southeast Asia, and is spreading rapidly in Africa and Papua New Guinea (Panta *et al.*, 2024).
- **Use of Snail**: *Pomacea canaliculata* (Lamarck snails) have shown promise in controlling aquatic weeds, including *Marisa cornuarietis* in Florida and *Anacharis alensa* in Brazil. It has also been reported that aquatic weeds such as *Ceratophyllum demersum*, *Najas guadalupensis*, and *Potamogeton illinoensis* can be successfully managed. The snail having root pruning activity significantly slowed down the growth and flowering of *Eichhornia crassipes*, while *Pistia stratiotes* and *Alternanthera philoxeroides* were only somewhat constrained (Archambault and Cope, 2016).

- **Chemical Control**

Choosing and using the right herbicides to eliminate aquatic weeds is essential for efficient management that preserves ecosystem health. Commercial fish ponds may be treated with herbicides to manage weed growth. Accurately identifying the problematic weed is the first step towards effective chemical control. 2,4-D The safe application of imazapyr, sodium carbonate peroxyhydrate compounds, triclopyr, copper, diquat, endothall, glyphosate, and carfentrazone ethyl in ponds has been permitted by the US EPA Herbicide 2,4-D is an effective solution for managing water hyacinth and smaller floating weeds such as Spirodela, Lemna, and Azolla. It is also effective in controlling Nymphaea and Nymphoides when used with copper sulphate pellets. Examples of submerged weeds in fish ponds include bladderwort, Hydrilla, Nechamandra, coontail, water plantain, and tape grass. Applying copper sulphate pellets locally using mud is an effective method. Additionally, 2,4-D amines and esters can be sprayed to eradicate rushes, sedges, and grasses.

Conclusion

The chapter discusses the abundance, roles, negative impacts, and management measures of aquatic weeds, with a special focus on the Kashmir valley. The water bodies in the valley host a variety of aquatic plants, some native and others introduced either accidentally or deliberately. These plants play crucial roles, such as providing natural food for the aquatic ecosystem and offering medicinal and consumptive value. However, the chapter also highlights how the overpopulation of aquatic weeds can disrupt the ecosystem balance. The primary aim is to provide valuable information on the current diversity of aquatic weeds in the Kashmir valley. The authors recommend that the scientific community and stakeholders maintain weed diversity at optimal levels and conduct periodic surveys to monitor the status of aquatic weeds. Immediate measures should be taken if invasive species become overabundant to ensure the health of the ecosystem in these water bodies over the Kashmir valley.

References

Aasim, M., Bakhsh, A., Sameeullah, M., Karataş, M. and Khawar, K.M., 2018. Aquatic plants as human food. Global perspectives on underutilized crops, pp.165-187.

Adedokun, M.A., Tairu, H.M., Adeosun, O. and Ajibola, O., 2017. Assessment of the optimal replacement levels of maize with water lettuce leaf (Pistia stratiotes) based diets for Clarias gariepinus.

Akwee, E.P., Netondo, G., Kataka, J.A. and Palapala, J.A., 2014. A critical review of the role of taro Colocasia esculenta L. (Schott) to food security: A comparative analysis of Kenya and Pacific Island taro germplasm.

Aladedunye, F., Przybylski, R., Rudzinska, M. and Klensporf-Pawlik, D., 2013. γ-Oryzanols of north American wild rice (Zizania palustris). Journal of the American Oil Chemists' Society, 90(8), pp.1101-1109.

Anand, A., Priyanka, U., Nayak, V.L., Zehra, A., Babu, K.S. and Tiwari, A.K., 2019. Nutritional composition and antioxidative stress properties in boiled tuberous rhizome of Neel Kamal (Nymphaea nouchali Burm. f.). Indian Journal of Natural Products and Resources (IJNPR [Formerly Natural Product Radiance (NPR)], 10(1), pp.59-67.

Archambault, J.M. and Cope, W.G., 2016. Life stage sensitivity of a freshwater snail to herbicides used in invasive aquatic weed control. Freshwater Mollusk Biology and Conservation, 19(2), pp.69-79.

Casanova, N.A., Ariagno, J.I., López Nigro, M.M., Mendeluk, G.R., Gette, M.D.L.A., Petenatti, E., Palaoro, L.A. and Carballo, M.A., 2013. In vivo antigenotoxic activity of watercress juice (Nasturtium officinale) against induced DNA damage. Journal of Applied Toxicology, 33(9), pp.880-885.

Ganie, A.H., Rasheed, S., Khuroo, A.A. and Dar, G.H., 2020. An updated checklist of aquatic macrophytes in Jammu and Kashmir state. Biodiversity of the Himalaya: Jammu and Kashmir State, pp.521-543.

Gorito, A.M., Ribeiro, A.R., Almeida, C.M.R. and Silva, A.M., 2017. A review on the application of constructed wetlands for the removal of priority substances and contaminants of emerging concern listed in recently launched EU legislation. Environmental pollution, 227, pp.428-443.

Hossain, M.E., Hillol Sikder, H.S., Kabir, M.H. and Sarma, S.M., 2015. Nutritive value of water hyacinth (Eichhornia crassipes).

Irabor, A.E., Obakanurhie, O., Nwachi, F.O., Ekokotu, P.A., Ekelemu, J.K., Awhefeada, O.K., Adeleke, L.M., Pierre Jrn, H. and Adagha, O., 2022. Duckweed (Lemna minor) meal as partial replacement for fish meal in catfish (Clarias gariepinus) juvenile diets. Livestock Research for Rural Development, 34(1), p.2022.

Kaul, V., Zutshi, D.P. and Vass, K.K., 1976, June. Aquatic weeds in Kashmir. In Aquatic weeds in South East Asia. Proceedings of regional seminar on noxious aquatic vegetation, New Delhi (pp. 79-83).

Kawade, S.S., Panchakarla, S. and Sapkale, P.H., 2023. Review on Aquatic Weeds and Their Management. Environment and Ecology, 41(4C), pp.2900-2908.

Khan, M.A., Shah, M.A., Mir, S.S. and Bashir, S., 2004. The environmental status of a Kashmir Himalayan wetland game reserve: Aquatic plant communities and eco-restoration measures. Lakes & Reservoirs: Research & Management, 9(2), pp.125-132.

Mangesh Kumar, M.K., Dhuria, R.K., Dinesh Jain, D.J., Sharma, T., Nehra, R. and Prajapat, U.K., 2018. A nutritional evaluation of Azolla (Azolla pinnata) as feed supplement.

Munibul, I., 2022. Nutritional Evaluation of Aquatic Weeds from Water Bodies of Central Kashmir (Doctoral dissertation, SKUAST Kashmir).

Nandi, S.K., Suma, A.Y., Rashid, A., Kabir, M.A., Goh, K.W., Abdul Kari, Z., Van Doan, H., Zakaria, N.N.A., Khoo, M.I. and Seong Wei, L. 2023. The potential of fermented water spinach meal as a fish meal replacement and the impacts on growth performance, reproduction, blood biochemistry and gut morphology of female stinging catfish (Heteropneustes fossilis). Life, 13(1), p.176.

Oelke, E.A., 1993. Wild rice: domestication of a native North American genus.

Panta, S., Schwarzländer, M., Weyl, P.S., Hinz, H.L., Winston, R.L., Eigenbrode, S.D., Harmon, B.L., Bacher, S. and Paynter, Q. 2024. Traits of insect herbivores and target weeds associated with greater biological weed control establishment and impact. BioControl, pp.1-16.

Reshi, Z., Shah, M.A., Rashid, I. and Khuroo, A.A. 2008. Alien plant invasions: Threat to wetland health and public wealth. Sarovar Saurabh, 4(1), pp.2-5.

Richardson, D.M., Pyšek, P., Rejmanek, M., Barbour, M.G., Panetta, F.D. and West, C.J. 2000. Naturalization and invasion of alien plants: concepts and definitions. Diversity and distributions, 6(2), pp.93-107.

Rose, P., Faulkner, K., Williamson, G. and Mithen, R. 2000. 7-Methylsulfinylheptyl and 8-methylsulfinyloctyl isothiocyanates from watercress are potent inducers of phase II enzymes. Carcinogenesis, 21(11), pp.1983-1988.

Safo-Kantanka, O., 2004. Colocasia esculenta (L.) Schott. Record from PROTA4U. PROTA (Plant Resources of Tropical Africa/Ressources Végétales de l'Afrique Tropicale).

Shah, M.A. and Reshi, Z.A. 2014. Characterization of alien aquatic flora of Kashmir Himalaya: implications for invasion management. Tropical Ecology, 55(2), pp.143-157.

Shah, M.A., Shah, A.B. and Reshi, Z.A. 2020. Aquatic ecosystems of Jammu and Kashmir State. Biodiversity of the Himalaya: Jammu and Kashmir State, pp.209-224.

Stadtlander, T., Förster, S., Rosskothen, D. and Leiber, F. 2019. Slurry-grown duckweed (Spirodela polyrhiza) as a means to recycle nitrogen into feed for rainbow trout fry. Journal of Cleaner Production, 228, pp.86-93.

Wasagu, R.S., Lawal, M., Shehu, S., Alfa, H.H. and Muhammad, C. 2013. Nutritive values, mineral and antioxidant properties of Pistia stratiotes (water lettuce). Nigerian Journal of Basic and Applied Sciences, 21(4), pp.253-257.

Yousif, R.A., Abdullah, O.J., Ahmed, A.M., Adam, M.I., Ahmed, F.A.M. and Idam, O.A., 2019. Effect of replacing fishmeal with water spinach (Ipomoea aquatica) on growth, feed conversion and carcass composition for Nile tilapia fry (Oreochromis niloticus). J Aquat Sci Mar Biol, 2(4), pp.3-20.

12

Introduction of Exotic Fishes in Kashmir Valley and Its Effects

Hudisa Banoo, Farooz Ahmad Bhat, Rinkesh Nemichand Wanjari, Zaheer Abass, Dechen Chuskit*, Mir Iqra Farooq and Arizo Jan

Division of Fisheries Resources Management, Faculty of Fisheries, SKUAST-K Rangil, Nagbal, Jammu and Kashmir, India

Abstract

Significant ecological, economic, and sociocultural interest has been shown in the introduction of exotic fish species to the Kashmir Valley. This chapter explores the methods, background, and effects of bringing non-native fish species into the aquatic environments of this area. Enhancing fishing resources, boosting local economies, and offering alternative livelihoods were the main goals. However, the intricate ecological interactions that have resulted from these invasions affect biodiversity, aquatic habitats, and native fish populations. The chapter describes the several exotic species that have been introduced, including the brown trout (Salmo trutta) and rainbow trout (Oncorhynchus mykiss), and it looks at how they have adapted to and multiplied in the valley's particular natural circumstances. A careful analysis of the relationships between these alien species and native aquatic life is presented, emphasizing instances of disease transfer, competition, and predation. The ecological effects on the local animals and plants are also investigated, including changes to food webs and habitat architecture. The economic effects are also covered, including how these alien species have affected regional fisheries, tourism, and subsistence fishing methods. Although the introduction of these species has produced benefits to the economy and expanded chances for recreational fishing, it has also created problems for the viability of local fish populations and traditional fishing communities.

Keywords: *Exotic fish species, Kashmir Valley, Ecological impact, Native biodiversity, Fisheries management, Non-native fish*

Introduction

An organism can be highly destructive in one habitat while coexisting peacefully in another. It is exceedingly challenging to forecast the impact of a new species on the local flora and fauna when it is brought into an ecosystem where it has never before lived naturally. In competition for resources and space, the imported species can surpass the native fauna. A delicate web of interdependence that gradually shaped over thousands of years is harmed when introduced species compete more successfully for resources because they have no predators in the area. In certain cases, they even kill and eat native populations (Raman *et al.*, 2013). Fish species classified as non-native are those that have been intentionally introduced outside of their natural range (for example, to support aquaculture, improve a reservoir or recreational fishery, through the ornamental fish trade, or to control mosquitoes), accidentally introduced (for example, through bait releases, aquaculture escapes, or ballast water transport (Gupta and Everard, 2019), or introduced in the name of well-meaning but misguided religious practices (Everard *et al.*, 2019).

Due to their higher birth rate, the tendency to feed on local fish species, increased growth rate, larger size, longer life spans, or competition for comparable food supplies and habitat, these species may negatively impact native fish species (Ricciardi and Rsmussen, 1998).

The International Union for Conservation of Nature and Natural Resources (IUCN) classifies alien invasive species as those that spread quickly through natural or semi-natural ecosystems or habitats, act as change agents, and endanger the biological diversity of the host species. These invasive are widely dispersed across the world's ecosystems, including every class of living thing (Raghubanshi *et al.*, 2005). Fish native to the area compete with exotics for food and habitat. The quality of the aquatic ecosystems may deteriorate, new illnesses and parasites may be introduced, hybrids may be produced, native fish may be killed, and genetic "pollution" may occur. All of these will eventually cause the priceless biodiversity to disappear (Nyman, 1991). The socioeconomic facets of the human population that depend on aquatic environments for survival are also at danger, in addition to possible consequences on the quantity or quality of biodiversity (Philipp *et al.*, 1995). The most varied group of vertebrates is fish, and freshwater fish in particular are thought to be among the most critically threatened animal species due to the introduction of alien (or non-native) species, which are species that do not typically occur in an aquatic environment (CONABIO, 2008). According to Contreras-Balderas (1999), these species hurt local populations when they are introduced and harm ecosystems by causing the water body to deteriorate. Maclvor brought the first group of exotic fish in 1874, namely crucian carp

(*Carassius carassius*) and tench (*Tinca tinca*), two cold-water species, introduced into Ooty Lake. Less success has been seen when attempting to transfer these fish into the plains. River pressure may be exacerbated by the ubiquity and dispersal of non-native fish species. According to Gupta and Everard (2019), these species have the potential to negatively impact native fish through a variety of means, including hybridization (the crossing of native and introduced fish), the spread of novel pathogens, adverse effects on the local food web and habitat structure, contaminant transfer, preferential growth in simplified hydrological regimes, and other competitive mechanisms. In addition to having direct competitive impacts, different life strategies adopted by native and introduced species (e.g., common carp tend to dig up sediment, whereas silver carp are filter-feeders) can alter the physical structure, water quality, and species balance of entire ecosystems. This is especially true when other human pressures are added.

History of Exotic Fish Introduction in the Kashmir

Enhancement of fisheries, aquaculture methods, and ornamental reasons are some of the driving forces for the introduction of exotic fish species in Kashmir and other parts of India (Jacob *et al.*, 2020). There have been major social, economic, and ecological effects of this process. Many exotic fish species were first introduced to the Indian subcontinent by the British colonial government. The main goals were to improve local fisheries and promote sport fishing. *Salto trutta fario*, or brown trout, and *Oncorhynchus mykiss,* or rainbow trout, were two of the most prominent and early arrivals to Kashmir. The British brought these species in the early 1900s to encourage fishing (Qadri and Yousuf, 1980). The focus on aquaculture increased after India's independence, which resulted in the introduction of more exotic species. Introduced for aquaculture, common carp (*Cyprinus carpio*) quickly rose to prominence as one of the most extensively produced species because of its adaptability and financial worth (Lone *et al.*, 2009). To reduce aquatic weeds and increase fish output, grass carp and silver carp were introduced in the 1960s and 1970s (Bhagat and Suresh, 2007). New species and better strains of previously imported fish have been introduced as a result of ongoing attempts to improve aquaculture. The introduction of decorative fish species for the aquarium commerce has increased (Bhagat and Suresh, 2007). Since trout were introduced, Kashmir has become a center for trout production in India, with the construction of many trout farms. In addition to meeting the needs of nearby markets, these farms also fulfill demand from other states and even export markets (Qadri and Yousuf, 1980). Dal Lake has been stocked with exotic species, such as grass carp and common carp. Although they have improved the output of fish, there are worries regarding their effects on the environment and native species of the lake (Lone *et al.*, 2009).

Benefits of Exotic Aquatic Species Introductions

The economics, ecology, and biodiversity of Kashmir have all benefited from the introduction of exotic species. Better fisheries productivity, more leisure time, and ecological balance through interactions between species are some of these advantages. An increased output of fisheries is one of the main advantages of bringing exotic aquatic species to Kashmir. Due to the huge rise in local fish production brought about by species like common carp (*Cyprinus carpio*) and rainbow trout (*Oncorhynchus mykiss*), local people now have a dependable source of protein and income (Mookerjee, 2020). According to Pandit *et al.* (2019), the introduction of these species has resulted in the growth of inland fisheries and aquaculture, improving food security and providing employment. The encouragement of leisure time activities is an additional important advantage. Kashmir has become a well-liked location for sport fishing due to the presence of exotic species like trout, drawing tourists and boosting the local economy (Sharma *et al.*, 2021). In addition to boosting local companies, the flood of visitors out fishing for fun also spreads awareness of conservation activities in the water (Gupta and Kumar, 2022). From an ecological point of view, invasive species can help keep some native species' numbers under control and the environment in balance. To preserve the general health of the water bodies and lessen the strain on aquatic plants, for example, the introduction of certain predatory fish can aid in managing the populations of smaller, potentially overabundant species (Singh and Singh, 2020). Furthermore, certain foreign species feed on algae and debris, which lowers eutrophication and encourages cleaner water (Kumar *et al.*, 2018). This helps to enhance the quality of the water. Furthermore, introducing resilient alien species can increase resistance to alterations in the environment and habitat deterioration. Pollution and climate change can be mitigated by species like common carp, which can flourish in a variety of environmental settings (Ahmad and Rather, 2017). Even in situations where native species may be having difficulties because of environmental pressures, this flexibility guarantees a steady supply of fish (Bhat and Yousuf, 2019). That being said, to prevent any possible harm to native ecosystems and biodiversity, foreign species introduction must be properly managed. Pandit and Kumar (2020) assert that the implementation of efficient management strategies and ongoing oversight are vital in ensuring that the advantages exceed the hazards.

1. **Pathway of introduction**

A. Artificial Introduction

The intentional act of introducing new species to certain habitats or returning existing species to places where they were formerly common is referred to as artificial introduction. According to Simberloff (2009), this is frequently done for financial gain or to restore the natural equilibrium. Fish species could be added to a body of water, for instance, to increase nearby fisheries or manage pest populations (Moyle and Marchetti, 2006). For the new species to be able to integrate without seriously disrupting the current ecosystem, a great deal of planning and management may be necessary (Crawford and Muir, 2008). The intentional introduction of Pacific salmon to North America's Great Lakes is a well-known example. This was done to create a sport fishery and manage the invasive alewife population (Johnson and Goettl, 2008). At first, the introduction was effective in keeping the alewives under control and offering chances for recreational fishing. But it also brought up intricate ecological relationships that require ongoing management (Crawford and Muir, 2008). Artificial introductions can benefit from several factors. According to Simberloff and Rejmánek (2011), they can boost biodiversity, supply fresh resources for regional economies, and aid in the restoration of ecological equilibrium. They do, however, also entail hazards. Introduced species have the potential to become invasive if they are not managed properly, outcompeting native species for resources and seriously disrupting the ecosystem (Ricciardi and Cohen, 2007). This emphasizes how crucial it is to prepare ahead and keep an eye on things while using artificial introductions (Simberloff and Rejmánek, 2011).

B. Occasional introduction

The unintended or accidental introduction of species into new environments is referred to as "occasional introduction" (Carlton, 2001). These invasions can have detrimental effects on nearby ecosystems and frequently happen without any previous preparation (Ruiz *et al.*, 2000). Numerous cases of accidental introductions have resulted from the unintentional transfer of species through international trade, travel, and transportation (Hulme, 2009). The introduction of the comb jellyfish, *Mnemiopsis leidyi*, to the Caspian Sea, is one such instance. Tankers' ballast water discharges probably caused the introduction of this species (Kideys, 2002). The arrival of Mnemiopsis has had a profound effect on the delicate environment of the Caspian Sea. The comb jellyfish feeds on zooplankton, which is an essential component of the food chain. Their overabundance has caused zooplankton numbers to fall, which has impacted fish populations that rely on them for sustenance. Significant efforts have been needed to control the ecological impact of this disturbance, which has had negative economic effects on the local fishing business (Shiganova, 1998). The introduction of the algae *Caulerpa taxifolia* to the Mediterranean Sea is another illustration. This incident happened when a Caulerpa specimen was dropped into the sea from a Monaco marine aquarium (Meinesz and Hesse, 1991). The algae outcompeted native marine vegetation and swiftly colonized wide stretches of the French and Italian coasts. According to Ceccherelli and Campo (2002), *C. taxifolia* covers the seafloor in dense mats that change ecosystems and decrease biodiversity. It is a prominent invasive species in the Mediterranean due to its quick expansion and difficulties in suppressing it (Piazzi and Ceccherelli, 2006). Occasional invasions can have a significant effect on regional ecosystems. Because they may not have natural predators in their new habitat, introduced animals may be free to spread uncontrolled. According to Bax *et al.* (2003), this may result in the eradication of native species, modifications to habitat architecture, and adjustments to the dynamics of regional food webs. Reduced biodiversity and changed ecosystem functioning are frequent outcomes of these changes, which can have cascade impacts across the ecosystem (Mack *et al.*, 2000).

Introductions in India

During the last many decades, more than 3000 species of exotic fishes have been introduced in India for aquaculture, sport fishing, ornamental fish keeping, mosquito control etc., besides a species of mollusc (*Mytilopsis sallei* native to Central America, introduced to Indian waters probably through ship fouling) (Raman *et al.*, 2013). Several exotic fishes are now well established in the natural water bodies of India. Although not many studies have been carried out on the impacts of exotic species in Indian waters, reports from the world over equivocally prove the highly deleterious impacts of exotic fishes in the Indian aquatic ecosystems. Some important fish introductions are listed in (Table 1).

Table 1. List of fish species that have been introduced to India (modified from Raman, *et al.*, 2013).

S.No.	Fish species	Year of introduction	Native country
A. Game fishes			
1.	Brown trout (*Salmo trutta fario*)	1863-1900	U.K.
2.	Loch Leven Trout (*Salmo levensis*)	1863	U.K.
3.	Rainbow Trout (*Salmo garirdnen*)	1907	Sri Lanka & Germany
4.	Eastern Brook Trout (*Salvelinus fontinalis*)	1911	U.K.

5.	Sockeye salmon (*Oncorhynchus nerka*)	1968	Japan
6.	Atlantic Salmon (*Salmo salar*)	1968	U.S.A.
B. Food Fishes			
1.	Golden Carp (*Carassius carassius*)	1870	U.K.
2.	Tench (*Tinca tinca*)	1870	U.K.
3.	Gourami (*Osphronemus goramy*)	1916	Java & Mauritius
4.	Common Carp (*Cyprinus carpio*) German strain	1939	Sri Lanka
5.	Tilapia (*Oreochromis mossambicus*)	1952	Africa
6.	Common Carp (*Cyprinus carpio*) Bangkok strain	1957	Thailand
7.	Grass Carp (*Ctenopahryngodon idella*)	1957	Japan
8.	Silver Carp (*Hypophthalmichthys molitrix*)	1959	Hong Kong
9.	Tawes (*Puntius javanicus*)	1972	Indonesia
10.	Sutchi (*Pangasianodon hypophthalmus*)	1997	Vietnam
11.	Whiteleg shrimp (*Litopenaeus vannamei*)	2010	
C. Larvicidal Fishes			
1.	Guppy (*Poecilia reticulata*)	1908	South America
2.	Top Minnow (*Gambusia affinis*)	1928	Italy
	D. Ornamental Fishes		
1.	Live bearers (27 species)	From various countries	
2.	Eggs layers (261 species)	From various countries	
E. Unauthorised introduction			
1.	Bighead carp (*Aristichthus nobilis*)		
2.	African Catfish (*Clarias gariepinus*)		
3.	Nile Tilapia (*Oreochromis niloticus*)		
4.	Red Tilapia (*Oreochromis* sp.)		
5.	Red Piranha (*Serrasalmus natteren*)		

Commonly Introduced Fish Species and their Origin in Kashmir

Improving aquaculture and fisheries became more of a priority after India gained independence in 1947. During this time, the common carp (*Cyprinus carpio*), a species native to Asia and Europe, was introduced. Because common carp are hardy and flexible in a range of environments, they were brought in and are perfect for aquaculture (Lone *et al.*, 2009). Other species, such silver carp (*Hypophthalmichthys molitrix*) and grass carp (*Ctenopharyngodon idella*), which are indigenous to the freshwater systems of East Asia, especially China, were brought in throughout the 1960s and 1970s. These species were introduced mainly to improve fish output in nearby water bodies and manage aquatic vegetation (Bhagat and Suresh, 2007). Due of its rapid growth and appropriateness for aquaculture methods in the Kashmir Valley, the mirror carp (*Cyprinus carpio* var. *specularis*), a type of common carp that was carefully developed, was also introduced. Because of its financial advantages, this species is becoming a mainstay in the local aquaculture industry (Lone *et al.*, 2009). The diversification of the aquatic fauna in the region has increased due to the introduction of different species brought in by the ornamental fish trade in recent years from South America, Africa, and Southeast Asia (Bhagat and Suresh, 2007). The effects of invasive species in Dal Lake provide for an interesting case study. To increase fish output, common and grass carp were introduced to this well-known body of water. Despite the fact that these species have flourished and benefited the local fisheries, there are worries about how they may affect the ecosystem of the lake, especially in light of the fact that they may displace native fish species and alter the water's quality (Lone *et al.*, 2009). Similarly, the introduction of trout has also resulted in the development of other trout farms around Kashmir, turning the area into a center for trout production in India. According to Qadri and Yousuf (1980), the aquaculture industry in the Kashmir Valley has the potential to provide significant economic benefits. These farms have not only catered to local markets but have also satisfied the need from outside markets. The introduction of exotic fish species in the Kashmir Valley illustrates the complex interplay between ecological impacts and economic benefits. While these introductions have enhanced local fisheries and aquaculture, they have also posed significant challenges to the region's native biodiversity. The ongoing management and monitoring of these species are crucial to ensuring a balance between economic development and ecological sustainability. Common exotic fish species of Kashmir Valley (Fig. 1).

Carassius carassius (Linnaeus, 1758)

Oncorhynchus mykiss (Walbaum, 1792)

Ctenopharyngodon idella (Valenciennes, 1844) *Cyprinus carpio Linnaeus,* 1758

Hypophthalmichthys molitrix (Valenciennes, 1844) *Salmo trutta Linnaeus,* 1758

Fig. 1 Common exotic fish species of Kashmir Valley.

Ecological Impact on Native fish Communities

Native fish communities in Kashmir have had major ecological effects as a result of the introduction of alien fish species into the region's water ecosystems. These effects come in many forms, such as disease introduction, habitat modification, predation, and competition for resources. Native fish are in competition with exotic species such as *Oncorhynchus mykiss* (rainbow trout) and *Cyprinus carpio* (common carp), which are able to outcompete them because of their aggressive nature and ability to adapt to a variety of environmental circumstances. Native species that are less competitive and develop more slowly, like the endemic *Schizothorax* species, may see a decline in population as a result of this competition. Moreover, the dynamics of the food web are upset by foreign species' predation, which lowers native biodiversity. Changes in water quality and sediment composition are two habitat adjustments brought about by alien fish, which put additional strain on local fish populations (Shafi *et al.*, 2020). Furthermore, the loss in native populations may be accelerated by alien species acting as vectors for diseases and parasites that were not previously prevalent in the local environment (Mir *et al.*, 2013).

Dal Lake: The introduction of common carp has resulted in notable ecological alterations to Kashmir's well-known lake, Dal Lake. During their foraging, these fish disrupt the sediments on the lakebed, which increases turbidity and releases nutrients that encourage the growth of algal blooms. The local fish species have suffered as a result, especially the *Schizothorax* sp., which depend on clean water for mating and eating. Research has indicated that the increase in common carp populations is correlated with a significant decrease in *Schizothorax* populations (Yousuf *et al.*, 2019).

Nigeen Lake: The introduction of rainbow trout to Nigeen Lake for aquaculture has put local fish species under intense competition. Adversarial predators, rainbow trout have been seen to feed on young native fish, such as *Schizothorax* sp. Native fish populations have significantly decreased as a result of this aggressive behavior and food competition. According to research by Qadri *et al.* (2015), once rainbow trout populations were established in the lake, there was a 40% decrease in the number of native fish species.

Wular Lake: The introduction of grass carp to Wular Lake, one of the biggest freshwater lakes in Asia, has caused ecological disruptions. The introduction of these herbivorous fish was intended to regulate the aquatic vegetation; but, because of their eating preferences, macrophytes which are essential to preserving the biological balance of the lake—have been overgrazed. A decrease in biodiversity has resulted from the removal of macrophytes, which has had a domino effect on the availability of habitat for native fish and invertebrates. According to Mir *et al.* (2013), native fish spawning habitats have been notably affected by the removal of plant cover, which has reduced reproductive success.

Socioeconomic Effect on Local Fishermen

Local fishermen in the Kashmir Valley have seen major socioeconomic effects as a result of the introduction of alien fish species. These effects are complex and affect the fishing communities' cultural customs, means of subsistence, and stability of income. Here's a thorough analysis of these consequences,

Impact on Livelihoods and Income

Decline in Native Fish Populations: There have been reports that the native *Schizothorax* sp. population in Dal Lake has been impacted by the introduction of *Cyprinus carpio* var. *specularis* (Singh and Lakra, 2011). Due to their historical preference in the local market, these natural species are essential to the local fisherman. The finding of Rather *et al.* (2016) shows how the rivalry between alien and native species for food and habitat has led to a decrease in native fish harvests, which has an impact on local fishermen's revenue.

Economic Competition: Compared to native species, exotic species frequently develop more quickly and are more robust. As a result, exotic fish now predominate in the neighbourhood markets and are occasionally offered for less money. It is hard for fishermen who catch mostly local species to compete on the economic front. Rashid *et al.* (2017) provide a case study from Dal Lake that shows how the introduction of less expensive exotic fish has resulted in a considerable decline in the profits of fishermen who target native species.

Changes in Fishing Practices

Adaptation to New Species: To capture alien species, local fishermen have had to modify their gear and fishing methods. This change is not always easy because it frequently necessitates spending money on new equipment and picking up new abilities. For example, research by Dar *et al.* (2015) in the Jhelum River basin demonstrates that to effectively catch the common carp, fishermen have to switch from using conventional fishing nets to more advanced ones.

Shift in Seasonal Patterns: The seasonal patterns of fishing have changed due to the introduction of foreign species. Compared to native species, exotic species may have distinct feeding and breeding cycles. Because of this, fishermen must modify their fishing schedules, which may cause disruptions to their customs and the seasonal economic planning they have practised for many years (Wani *et al.*, 2018).

Sociocultural Impacts:

Erosion of Cultural Heritage: In Kashmir, fishing is not just a commercial endeavor but also a cultural custom. The loss of traditional fishing knowledge and techniques that have been passed down through the generations is a result of the dominance of foreign fish species over local species. A research by Mir and Pandit (2015), which records the decrease in the usage of traditional fishing methods among newer generations of fishermen, highlights this destruction of culture.

Community Displacement: Fishermen have occasionally been forced to migrate as a result of overfishing and environmental deterioration brought on by the introduction of alien species. Social networks and community structures are upset by this relocation. According to a case study from Wular Lake (Kumar and Pandit, 2016), dwindling fish populations in their home waters have forced several fishing tribes to relocate to metropolitan regions in search of other means of subsistence.

Environmental Degradation and Its Socioeconomic Effects

Habitat Degradation: Fish populations are further impacted by habitat deterioration that is frequently brought about by the introduction of alien species. For instance, it is well known that when feeding, common carp may agitate sediments, increasing turbidity and destroying aquatic vegetation. The general health of the aquatic ecosystem is impacted by this environmental deterioration, which lowers productivity and, as a result, fish capture (Rashid *et al.*, 2017).

Increased Costs for Sustainability: The management and reduction of the effects of invasive species can need large funding commitments to conservation and restoration initiatives. Local fishing communities may bear these expenses directly or indirectly via modifications to fishing laws and practices. According to research by Ahmad and Mir (2018), local fishermen's already fragile financial status has been further burdened by having to participate in community-led projects aiming at restoring native fish habitats. The economic effects on local fishermen of the introduction of exotic fish species in the Kashmir Valley are significant. The difficulties are manifold and intricate, ranging from falling native fish populations and modified fishing techniques to cultural loss and environmental deterioration. A multimodal strategy including community involvement, sustainable fishing management, and cultural heritage preservation is needed to address these problems.

Invasive Species Management And Control

Prevention and Early Detection

Putting strict controls in place for the release and importation of non-native animals. Before any new species are introduced, this entails careful screening and risk evaluations (Simberloff and Rejmánek, 2011). Supporting safe behaviors among fishers, aquarium enthusiasts, and the aquaculture business, as well as educating the public about the dangers posed by invasive species (Hulme, 2009). Putting in place monitoring systems to find fresh incursions quickly. To detect the presence of invasive species before they get established, this may include conducting routine surveys of water bodies and utilizing environmental DNA (eDNA) methods (Bax *et al.*, 2003).

Eradication and Control

Reducing the number of invasive species can be accomplished physically via netting, electrofishing, or targeted fishing. For instance, specialized nets are used to collect and remove common carp as part of the attempts to eradicate these species from Dal Lake (Yousuf *et al.*, 2019). Controlling the populations of invasive species can be achieved by introducing natural predators, parasites, or illnesses that target them particularly. To prevent unforeseen ecological repercussions, this strategy must be applied carefully (Mack *et al.*, 2000). Fish poisons known as piscicides can be used to eradicate invasive fish from smaller bodies of water. To reduce damage to the ecosystem and native species, this approach must be properly maintained (Shafi *et al.*, 2020).

Restoration and Rehabilitation

Restoring environments that invading species have damaged. To sustain native species, this may entail replacing native aquatic vegetation, reestablishing natural water flow patterns, and enhancing water quality (Mir *et al.*, 2013). Replenishing native fish species that have been displaced or supplanted by invasive species in water bodies. In doing so, native populations may be reestablished and ecological equilibrium may be restored (Bhagat and Suresh, 2007).

Integrated Management Approaches

Combining several management techniques for more efficient control of invasive species. Biological, mechanical, and chemical techniques may all be used in IPM strategies, depending on the needs of certain species and habitats (Simberloff and Rejmanek, 2011). Including government organizations, fishers, local residents, and conservation groups in the management process. Working together is essential to creating and carrying out successful management strategies (Ahmad and Mir, 2018).

Research and Monitoring

Carrying out studies to support management choices on the biology, ecology, and effects of invasive species. This entails researching their life cycles, methods of reproduction, and relationships with local species (Gupta and Everard, 2019) and putting in place extensive monitoring systems to keep tabs on the numbers of both native and alien species. According to Yousuf *et al.* (2019), this information is crucial for determining the success of management interventions and changing tactics as necessary.

Conclusion

Introducing exotic fish species in the Kashmir Valley has profoundly impacted the region's aquatic ecosystems, native biodiversity, and socio-economic structures. Although the initial goal was to enhance fishing resources and support local economies, the long-term ecological impacts have been significant. Species such as brown trout (*Salmo trutta*) and rainbow trout (*Oncorhynchus mykiss*) have adapted well to the Valley's environment, leading to their proliferation. However, their presence has disrupted local ecosystems by altering food webs, habitat structures, and native species populations. These exotic species compete for resources, prey on native fish, and spread diseases, threatening the survival of indigenous aquatic life. While economic benefits such as enhanced fisheries and increased tourism are notable, they come with significant ecological costs. The balance between economic gains and ecological health is delicate, with the sustainability of local fish populations and traditional fishing practices at risk. The Kashmir Valley's experience highlights the complexity of introducing exotic species. Predicting their impacts on local flora and fauna is challenging, as they often outcompete native species due to a lack of natural predators. This disruption can lead to the decline or extinction of native species and biodiversity loss. Sustainable fisheries management practices are crucial to mitigate these impacts and ensure that economic benefits do not come at the expense of the environment. Continued research and monitoring are essential to understand the long-term consequences of these introductions and develop strategies that balance economic development with ecological preservation.

References

Ahmad, I., & Mir, J. I. (2018). Community-led initiatives for conservation of native fish species in Kashmir Valley. Journal of Applied Ichthyology, 34(5), 987-994.

Ahmad, S., & Rather, M. A. (2017). Impact of Exotic Fish Species on Indigenous Fish Fauna of Kashmir Valley. Journal of Aquatic Ecology 25(3), 45-58.

Bax, N., Williamson, A., Aguero, M., Gonzalez, E., & Geeves, W. (2003). Marine invasive alien species: A threat to global biodiversity. Marine Policy, 27(4), 313-323.

Bhagat, R. C., & Suresh, B. (2007). The impact of introduced species on Kashmir's aquatic ecosystems. Aquatic Ecology, 41(1), 41-50.

Bhat, F. A., & Yousuf, A. R. (2019). Ecological and Economic Impact of Exotic Fish Species in the Kashmir Valley. Fisheries and Aquaculture Journal, 10(2), 109-115.

Carlton, J. T. (2001). Introduced species in US coastal waters: environmental impacts and management priorities. Pew Oceans Commission.

Ceccherelli, G., & Campo, D. (2002). Different effects of Caulerpa racemosa on two co-occurring seagrasses in the Mediterranean. Botanica Marina, 45(1), 71-76.

CONABIO (2008). Capital Natural de México, Vol I: Conocimiento actual de la Biodiversidad. México: CONABIO. 809 pp.

Contreras-Balderas, S. (1999). Annotated checklist of introduced invasive fishes in México, with examples of some recent introductions. In R. Claudi & J. H. Leach (Eds.), Nonindigenous freshwater organisms: Vectors, biology, and impacts (pp. 31-52). Boca Raton, FL: Lewis Publishers.

Crawford, S. S., & Muir, A. M. (2008). Global introductions of salmon and trout in the genus Oncorhynchus: 1870-2007. Reviews in Fish Biology and Fisheries, 18(4), 313-344.

Dar, S. A., Mir, J. I., Pandit, A. K., Rather, G. A., & Dar, M. A. (2015). Adaptation of fishing practices in response to invasive species in the Jhelum River basin. Fisheries Management and Ecology, 22(1), 73-82.

Dar, S. A., Pandit, A. K., & Rather, M. I. (2015). Adaptation strategies of fishermen to exotic fish species in Jhelum River Basin, Kashmir. Fisheries Research, 174, 108-115.

Everard, M., Pinder, A. C., Raghavan, R., & Kataria, G. (2019). Are wellintended Buddhist practices an under-appreciated threat to global aquatic biodiversity? Aquatic Conservation: Marine and Freshwater Ecosystems, 29, 136–141

Gupta, N., & Everard, M. (2019). Non-native fish introductions in India: their impact and management. Biodiversity and Conservation, 28(1), 23-39.

Gupta, R., & Kumar, N. (2022). Recreational Fishing and Tourism in Kashmir: A Case Study of Trout Fishing. Journal of Sustainable Tourism, 30(4), 550-565.

Hulme, P. E. (2009). Trade, transport and trouble: managing invasive species pathways in an era of globalization. Journal of Applied Ecology, 46(1), 10-18.

Jacob, T. C., Suresh, V. R., Sugunan, V. V., & Meenakumari, B. (2020). Conservation of aquatic biodiversity for achieving India's food, nutritional and livelihood security with special reference to inland fisheries and aquaculture. Journal of Fisheries Research, 4(1), 31-36.

Johnson, B. M., & Goettl, J. P. (2008). Pacific salmon in the Great Lakes: lessons from 40 years of stocking. Fisheries, 33(11), 511-524.

Kideys, A. E. (2002). Fall and rise of the Black Sea ecosystem. Science, 297(5586), 1482-1484.

Kumar, P., Singh, J., & Koul, P. (2018). Role of Exotic Fish Species in Aquatic Ecosystem Management in Kashmir. Environmental Management and Sustainability, 7(2), 112-121.

Kumar, S., & Pandit, A. K. (2016). Displacement and socio-economic impact of exotic fish species on fishing communities in Wular Lake, Kashmir. Environmental Science & Policy, 64, 101-110.

Lone, F. A., Rashid, I., Bhat, G. A., Bhat, R., & Ahmed, K. (2009). The status and impact of common carp (Cyprinus carpio) in Dal Lake, Kashmir. Journal of Aquatic Ecosystem Health & Management, 12(2), 97-104.

Lone, F.A., Balkhi, M.H., & Shah, T.H. (2009). Studies on fish diversity in Manasbal Lake, Kashmir. Journal of Himalayan Ecology and Sustainable Development.

Mack, R. N., Simberloff, D., Lonsdale, W. M., Evans, H., Clout, M., & Bazzaz, F. A. (2000). Biotic invasions: causes, epidemiology, global consequences, and control. Ecological Applications, 10(3), 689-710.

Meinesz, A., & Hesse, B. (1991). Introduction of the tropical alga Caulerpa taxifolia and its invasion of the northwestern Mediterranean. Oceanologica Acta, 14(4), 415-426.

Mir, F. A., & Pandit, A. K. (2015). Cultural impacts of exotic fish species on traditional fishing practices in Kashmir Valley. Asian Fisheries Science, 28(1), 30-40.

Mir, J. I., Chandra, S., & Sheikh, I. (2013). Ecological Impacts of Exotic Fish Species on the Fish Fauna of Kashmir Himalaya, India. Journal of Biodiversity and Environmental Sciences, 3(1), 17-24.

Mir, S. A., & Pandit, A. K. (2015). Cultural heritage and the impact of invasive fish species in Kashmir. Environmental Monitoring and Assessment, 187(7), 438.

Mookerjee, D. (2020). Fisheries Development through Introduction of Exotic Species in Kashmir. Journal of Fish Biology, 76(1), 101-110.

Moyle, P. B., & Marchetti, M. P. (2006). Predicting invasion success: freshwater fishes in California as a model. Bioscience, 56(6), 515-524.

Nyman, L. (1991) Conservation of freshwater fish. Protection of biodiversity and genetic variability in aquatic ecosystems. Fisheries Development Series 56, Swedmar & WWF, Sweden, 38 p.

Pandit, A. K., & Kumar, R. (2020). Managing Exotic Species in Kashmir's Aquatic Ecosystems. Biodiversity Conservation Journal, 14(1), 35-50.

Philipp, D.P., J.M. Epifanio, J.E. Mardsen and J.E. Claussen (eds.) (1995). Protection of Aquatic Biodiversity. Proceedings of the World F isheries Congress. Theme3, Oxford and IBH Publ. Co., New Delhi, 282 p

Piazzi, L., & Ceccherelli, G. (2006). Persistence of biological invaders in marine habitats. Marine Pollution Bulletin, 52(8), 866-871.

Qadri, M. Y., & Yousuf, A. R. (1980). Trout fisheries in Kashmir: history and impact. Fisheries Science, 46(4), 543-552.

Qadri, M. Y., Shafi, S., & Yousuf, A. R. (2015). Impact of Rainbow Trout Introduction on Native Fish Populations in Nigeen Lake. Aquatic Ecology Journal, 52(4), 301-310.

Qadri, M.Y., & Yousuf, A.R. (1980). Impact of Introduced Fish Species on the Native Fish Fauna of Kashmir. Proceedings of the Indian National Science Academy.

Raghubanshi, A.S., L.C. Rai, J. P. Gaur and J.S. Singh (2005) Invasive alien species and biodiversity in India, Current Science, 88 (4) : 539540.

Raman, R. P., Mishra, A., Kumar, S., Sahay, S., Bhagat, M. N., & Kumar, S. (2013). Introductions of exotic fish species into Indian waters: An overview of benefits, impacts. Advances in fish research, 6, 1-14.

Rashid, R., Dar, S. A., & Sofi, A. H. (2017). Socio-economic impact of exotic fish species on fishermen community of Dal Lake, Kashmir. Asian Journal of Agricultural Extension, Economics & Sociology, 19(2), 1-9.

Rashid, R., Khan, M. A., Pandit, A. K., & Dar, M. A. (2017). Environmental degradation due to exotic fish species in Dal Lake. Journal of Environmental Science and Health, 52 (8), 1-12.

Rather, M. A., Dar, M. A., Pandit, A. K., & Khan, M. A. (2016). The socioeconomic impact of declining native fish populations in Kashmir. Economic Botany, 70(4), 312-324.

Rather, M. I., Bhat, F. A., Yousuf, A. R., & Balkhi, M. H. (2016). Impact of exotic fish species on the native fish diversity in Dal Lake, Kashmir. Journal of Environmental Biology, 37(4), 775-780.

Ricciardi, A., & Cohen, J. (2007). The invasiveness of an introduced species does not predict its impact. Biological Invasions, 9(3), 309-315.

Ricciardi, A., & Rasmussen, J. B. (1998). Predicting the identity and impact of future biological invaders: A priority for aquatic resource management. Canadian Journal of Fisheries and Aquatic Sciences, 55, 1759–1765.

Ruiz, G. M., Carlton, J. T., Grosholz, E. D., & Hines, A. H. (2000). Invasion of coastal marine communities in North America: apparent patterns, processes, and biases. Annual Review of Ecology and Systematics, 31, 481-531.

Shafi, S. A., Khan, M. A., Rather, M. A., & Pandit, A. K. (2020). Ecological impact of invasive fish species in Kashmir. Aquatic Conservation: Marine and Freshwater Ecosystems, 30 (3), 467-482.

Shafi, S., Yousuf, A. R., & Qadri, M. Y. (2020). Impact of Exotic Fish Species on Native Fish Diversity of Kashmir. Journal of Fish Biology, 76(3), 535-553.

Sharma, R., Singh, A., & Verma, P. (2021). Tourism and Angling: The Economic Impact of Trout in Kashmir. Tourism Economics, 27(3), 523-537.

Shiganova, T. A. (1998). Invasion of the Black Sea by the ctenophore Mnemiopsis leidyi and recent changes in pelagic community structure. Fisheries Oceanography, 7(3-4), 305-310.

Simberloff, D., & Rejmánek, M. (2011). Encyclopedia of biological invasions. University of California Press.

Singh, A.K., Lakra, W.S. (2011). Risk and benefit assessment of alien fish species of the aquaculture and aquarium trade into India. Reviews in Aquaculture, 3, 3–18.

Singh, R., & Singh, H. (2020). Ecological Benefits of Exotic Fish Species in Inland Waters of Kashmir. International Journal of Ecology, 2020, Article ID 546789.

Wani, I. A., Rashid, R., & Bhat, F. A. (2018). Seasonal variations in fish catches due to exotic fish species in Wular Lake, Kashmir. Journal of Aquatic Biology & Fisheries, 6(2), 45-52.

Wani, M. A., Shafi, S. A., Khan, M. A., & Rather, M. A. (2018). Changes in seasonal fishing patterns due to invasive species in Kashmir. Environmental Biology of Fishes, 101 (4), 1-10.

Yousuf, A. R., Pandit, A. K., & Qadri, M. Y. (2019). Ecological Implications of the Introduction of Common Carp in Dal Lake, Kashmir. Ecological Studies, 45(2), 233-245.

Yousuf, A. R., Pandit, A. K., Khan, M. A., & Rather, M. A. (2019). Impact of common carp on Dal Lake's native fish species. Hydrobiologia, 846 (1), 125-138.

13

Exploration of Sports Fisheries Development, Techniques, Socio-Economic Benefits, and Sustainable Practices in Kashmir Valley

Manabjyoti Barman[1*], Sagar Ronad[1], Zaheer Abass[2], and Vipul Singh Badguzar[3]

[1]*ICAR-Central Institute of Fisheries Education, Panch Marg, off Yari Road, Andheri (West), Mumbai-400061 Maharashtra*

[2]*Faculty of Fisheries, Sher-e-Kashmir University of Agricultural Science and Technology of Kashmir, Shalimar, Srinagar, J&K UT, 190006*

[3]*MFSc Aquatic Environmental Management, ICAR-Central Institute of Fisheries Education, Panch Marg, off Yari Road, Andheri (West), Mumbai-400061 Maharashtra*

Abstract

This chapter explores the status, challenges, and opportunities of recreational fisheries in India, with a special focus on the Kashmir Valley. Sport fishing, or recreational fishing, is a popular activity that involves catching fish for pleasure or competition rather than for commercial or subsistence purposes. Historically rooted in India since ancient times, sport fishing has gained momentum, particularly in the colder regions of the North-East and the Himalayan states. The Kashmir Valley, often dubbed the "Angler's Paradise," offers abundant opportunities for sport fishing, especially with the introduction and establishment of species like brown trout and rainbow trout. The chapter highlights key game fish species, including the mighty mahseer, trout, and various indigenous fish, discussing their habitats and significance. It also addresses the regulatory landscape, emphasizing the need for sustainable practices and improved management to protect aquatic ecosystems. Challenges such as the lack of scientific data, unregulated stocking, and governance issues are identified, alongside the potential for sustainable development and ecotourism. There is a need for enhanced research, policy intervention, and community engagement to ensure the long-term viability and growth of recreational fisheries in India.

Keywords: *Recreational fishing, Kashmir Valley, Mahseer, Sustainable development*

Overview of Recreational Fisheries in India with Special Emphasis on Kashmir Valley

Sport fishing, often referred to as Recreational fishing is fishing for fun or competition. It differs from commercial or subsistence fishing which targets for profit or fishing for survival respectively (Mandal *et al.*, 2016). As per FAO (2012), recreational fishing is fishing of aquatic animals (mainly fish) that do not constitute the individual's primary resource to meet basic nutritional needs and are not generally sold or otherwise traded on export, domestic or black markets. It is one of the most adventurous things in the world. In the simplest form, sport fishing involves using rods, lines, hooks and bait or lures to attract and catch the fish (Mandal *et al.*, 2016). The fighting attributes of the target species leads to the sport of the fishermen in Sport fishing. Even though the sport fishing seems to be limited in the freshwater ecosystems it is extended to the marine waters also. Sport fishing involves catching of fishes which require skills as well as knowledge regarding the behaviour and food preferences of the target species. Sport or game fish are the targeted species that are caught by rods and lines and not released back into the water by the hobbyist or fishermen in disgust. However, releasing the catch to the environment should be a prerequisite for sport fishing considering the ecological and conservation point of view. In sport fishing, the target fish is played by the hobbyist until the fish is landed or lost (IGFA, 2016). According to Mandal *et al.* (2016), sport fish can be categorised into 2 group namely Small and Big. The fish which as less than 50 kg are grouped as the small ones like mahsser, trout etc. whereas fish weighing more than 50 kg are grouped as big ones like Goonch catfish.

Fishing is an old age practice and fishing by hooks and line is the primitive form. Fishing purely for fun and releasing caught fish is a much more modern development. Although it is not possible to clearly state when fishing was first said to be recreational, the book titled 'The Complete Angler' by Izzak Walton in 1963 described recreational fishing for the first time. In India, King Somesvara's (1127 AD) has mentioned about recreational fisheries (Matsyavinoda) in his treatise "Manasollosa" (Hora, 1951). With the record and description of mahseer from Ganges in the early 19th century, soon it became popular as an angling species through the efforts of the Oriental Sporting Magazine (Nautiyal, 2014). In the meantime, many different authors have pointed out about angling in India and highlighted about the sporting fishes.

British anglers have introduced first-time brown trout (1860s) and rainbow trout (1909) in streams and rivers of the Himalayas and Western Ghats (Sehgal, 1999a, b). Although recreational angling struggled to maintain its popularity after India's independence, the interests of both foreign and Indian anglers began to focus on Indian freshwater systems and their fish species in the 1970s. Established and emerging angling organizations across the country invested both time and money to build on the earlier foundations of sport fishing (Gupta *et al.*, 2015). The All India Game Fishing Association acting as a pivotal role in popularizing sport fishing in India with its state and regional representatives. It is a non-profit making, charitable organization with regional affiliates like Mustad, Maharashtra State Angling Association (MSAA)-Powai (Maharashtra), SHIMANO, The Rogue Angler, Anglers Association (Futala Lake)-Nagpur (Maharashtra), Gypsy's Camp-Bhakhra (Himachal) and Dasvino Town & Country Club.

In India, the North eastern states along with the upper Himalayan colder regions holds an immense potential for the development of sport fisheries. In the state of Arunachal Pradesh, the Subansiri, Siang, Lohit, Pare and Kameng River are the important river belt that sustains the big fighting fishes and is considered as one of the favourite spot for angling. Similarly, Jia Bhoroli and Manas Rivers in Assam; Dhansiri and Intangki, Milak and Dikhu, Doyang, Arachu, Lanyi, Seidzu, Tesuru and Tizu, Zungki and Likhimro in Nagaland; Teesta River and Rangeet in Sikkim are the major angling spots of India (Baruah and Sharma, 2018). The Kashmir valley has been referred to as the Anglers paradise, which supports a range of rivers and streams as well as high-altitude lakes which serve as the favourite angling location among the anglers. The river Indus and Liddar is characterised by the swollen water due to melting of glacier which provides the wild habitat for the mighty mahseers and the trouts. Angling has successfully been establishes as business in Kashmir during the months of spring and summer. The introduction of Brown trout and rainbow trout to the streams of Kashmir has been proved to be one of the successful introduction, where they thrived and has established the populations (Mahanata *et al.*, 2009). Among the sport fishes, the mahseers have exhibited a major socio-economic and recreation significance among the anglers in Kashmir Valley. Areas like the Tawi River (Jammu), Ujh Stream, Ravi River (Kathua), Tawi River (Udhampur), Poonch River (Poonch), Rajouri Wali Tawi (Rajouri), Salal Reservoir, Nallah Rud, Ans, Chenab, Pouni Nallah (Reasi) are the favourite angling locations characterised by the rocky boulder and torrent streams (Baruah and Sharma, 2018). Efforts should be made for the development of recreational fishing in the Kashmir valley along with regulatory interventions to sustain the long-term viability of sport fisheries.

Important Game Fish Species and Their Habitats

India has a rich aquatic diversity owing to its unique climatic and topographical conditions. It contributes nearly 27% to world diversity in terms of the native aquatic fauna, followed by China. Chandra *et al.* (2017) reported 3437 fish species from India, out of which 1027 have been reported from India. Considering sport fishing, many native as well as introduced fish species in India have been targeted by different anglers from all over the World due to the fighting nature of these species. In India, sport fishing has been practiced since long times in the North Eastern region and upper Himalayan states along with some parts of Southern India. Among the sport fishes, Mahseer (*Tor* sp.), Trout, and many Indigenous fish such as goonch catfish, snakeheads, and carps as well as eels have widely been targeted by the anglers. The major species targeted by recreational anglers in India include *Tor* sp. (previously known as *Tor mussullah* or the hump-backed mahseer), *T. putitora* (golden mahseer), *T. khudree* (Deccan mahseer), *Neolissochilus hexagonolepis* (copper or chocolate mahseer), and Gibelion catla (Indian major carp/catla) whereas native fish species such as *Channa diplogramma* (giant snakehead), C. marulius (bullseye snakehead), C. punctata (spotted snakehead), *C. striata* (snakehead murrel), *Wallago attu* (wallago), *Hemibagrus maydelli* (Krishna red tail Catfish), *Bagarius bagarius* (goonch), *Labeo rohita* (rohu), *Cirrhinus cirrhosus* (mrigal), *Hypophthalmichthys molitrix* (silver carp), *Cyprinus carpio* (common carp), *Ctenopharyngodon idella* (grass carp) and *Hypophthalmichthys nobilis* (big head carp) constitutes the occasional catch (Gupta *et al.*, 2015).

Mahseer, known as the undisputed king of Indian freshwaters is a large-bodied potamodrous fish with considerable economic and recreational interest. It is the most popular game fish among the anglers in India. Mahseer belongs to the family Cyprinidae and constitutes nearly 15 species in India belonging to 3 genera namely *Tor* (Gray), *Neolissochilus* (McClelland) and *Naziritor* (McClelland), out of 47 global species (Baruah and Sharma, 2018). Kulkarni and Ogale (1979) reported mahseer to be the noblest game fish in India. Owing to the toughest fighting nature of mahseer, it has been considered to be the favourite game fish both in India and overseas. Distribution of Mahseer species in different states of India are presented in (Table 1).

Table 1: Distribution of Mahseer species in different states of India (Baruah and Sharma, 2018)

State	River	Major species of mahseer
Arunachal Pradesh	Subansiri	*Tor putitora* and *Neolissochilus* sp.
	Siang and tributaries	*Tor putitora* and *Neolissochilus* sp.
	Pare	*Tor putitora* and *Neolissochilus* sp.
	Kameng	*Tor putitora* and *Neolissochilus* sp.
Assam	Jia Bhorelli	*Tor tor*, *T. putitora*, *T. mosal*, *T. progenius* and *Neolissochilus hexagonolepis*
	Manas	*T. putitora*, *T. tor* and *N. hexagonolepis*

Nagaland	Dhansiri, Intangki, Milak, Dikhu, Doyang, Arachu, Lanyi, Seidzu, Tesuru, Tizu, Zungki, Likhimro	*T. putitora* and *Neolissochilus* sp.
Sikkim	Teesta and Rangeet	*T. putitora* and *Neolissochilus* sp.
Meghalaya	Pools and sanctuaries	*Neolissochilus* sp.
Mizoram	Rivers of Kolodyne drainage	*Neolissochilus* sp.
Uttarakhand	Mahakali, Saryu	*T. putitora*
Peninsular India	Narmada, Cauvery	*T. khudree*

Trout is one of the most popular game fish species of India. In India, nearly 12 native species of *Schizothorax* (Snow trout) is found, of which *Schizothorax kumaonensis* is endemic to the country (Froese and Pauly, 2024). Moreover, the successful introduction of rainbow trout and Brown trout in the 90's in India has led to the development of recreational fishing in India, especially in the northern region of the country. Table 2 depicts important game fish species of India.

Table 2: Important game fish species of India (Gupta *et al.*, 2015)

Sr. No.	Scientific name	Common name
1.	*Anguilla bengalensis*	Indian mottled eel
2.	*Bagarius bagarius*	Goonch
3.	*Channa striata*	Striped or Cheveron snakehead
4.	*Cirrhinus cirrhosus*	Mrigal
5.	*Clarias gariepinus*	African sharp tooth catfish
6.	*Gibelion catla*	Catla
7.	*Hypophthalmichthys molitrix*	Silver carp
8.	*Labeo calbasu*	Orangefin labeo
9.	*Labeo rohita*	Rohu
10.	*Neolissochilus hexagonolepis*	Copper/Chocolate mahseer
11.	*Oncorhynchus mykiss*	Rainbow trout
12.	*Salmo trutta*	Brown trout
13.	*Schizothorax progastus*	Dinnawah snow trout
14.	*Tor khudree*	Deccan mahseer
15.	*Tor putitora*	Golden mahseer
16.	*Wallago attu*	Mully catfish/Freshwater shark

Commonly used Fishing Techniques

Bait-Casting

Bait-casting, an age-old fishing technique, relies on either live or artificial bait to entice fish, requiring skill and precision. It's favored by experienced anglers who enjoy its challenges. Typically, this method involves a reel mounted on top of a fishing rod, known for its precision and control, which aids in accurate bait placement. The reels in bait-casting setups are meticulously crafted for accuracy and control, allowing anglers to cast their bait precisely. The rods are sturdy and durable, capable of handling heavy loads and providing ample leverage for casting. This gear is versatile and suitable for targeting a wide variety of fish in both freshwater and saltwater environments. Rigging and presenting bait are crucial techniques, requiring anglers to carefully prepare their bait to effectively attract fish, whether it's live or dead, whole or cut into pieces, and presented on a hook or combined with artificial lures. Mastering casting accuracy is essential in bait-casting, as it involves controlling the release of a line from the reel and positioning the bait accurately near potential fish-holding areas, often using various bait rigs comprising hooks, leaders, swivels, and sinkers.

Spin-Casting

Spin-casting is recognized as one of the easiest fishing techniques, particularly suitable for beginners and kids due to its simple operation. This method involves using a spinning reel mounted beneath a fishing rod, which features a convenient push-button for casting. Spin-casting reels are known for their simplicity, requiring minimal skill to operate, making them ideal for newcomers to fishing. Accompanying these reels are lightweight and easy-to-handle spin-casting rods, allowing for effortless casting of lightweight lures and bait. The versatility of spin-casting equipment adds to its appeal, enabling anglers to fish in various environments, from calm freshwater lakes to dynamic coastal waters. Regarding technique, spin-casting excels in its simplicity, with the push-button mechanism making casting easy and requiring minimal effort. Additionally, spin-casting gear is effective at presenting a range of lures, including spinners, crankbaits, and soft plastics, enticing fish to bite. Due to its user-friendly nature, spin-casting is often recommended as the preferred method for beginners looking to start their fishing journey

Side Arm Casting

The side-arm cast technique starts with positioning the casting hand at belt level, with the rod parallel to the water or ground surface and aimed directly at your target. For right-handed casters, the left foot should be slightly ahead of the right, with the body angled towards the target. Hold the rod and reel with your right hand, fully depressing the release button with your thumb. The lure or plug should hang about 10 cm (4 in.) from the rod tip. With a swift motion, bring the rod to your right while keeping it parallel to the water surface. Stop the backward movement when the rod reaches a 90-degree angle from its starting position. As the casting plug's momentum bends the rod deeply, forcefully move the rod forward using both your forearm and wrist for maximum power. Release the thumb button just before your control thumb points at the target. If the plug veers to the right of the target, the button is released too early; if it goes left, it is released too late. It's advisable to demonstrate this cast to your group. Since the rod is held to the side rather than overhead, more space is needed per person during practice. Inform the group that the next session will include a casting game and contest, emphasizing both accuracy and proper technique (Henderson, 2000).

Spinning

Spinning, positioned between bait-casting and spin-casting in complexity, is a popular fishing method known for its adaptability and moderate skill level. This technique involves using a spinning reel attached beneath the fishing rod, available in both open-face and closed-face designs. Renowned for its versatility, spinning gear is well-suited for a wide range of fishing situations, from freshwater trout to saltwater gamefish. Spinning rods, complementing the reel, are longer and lighter than bait-casting rods, making them ideal for casting lightweight lures and bait over longer distances. The key feature of spinning gear is its user-friendly design and ability to handle light lines, making it perfect for finesse fishing approaches. In terms of technique, spinning gear excels in casting ability, allowing anglers to cast lightweight lures and bait far distances, reaching remote fish-holding areas inaccessible by other tackle types. Additionally, spinning tackle's versatility is evident in its effectiveness in presenting various lures, including jigs, spinners, and soft plastics, enticing fish with a diverse range of offerings. Spinning gear is particularly essential for finesse fishing techniques that rely on delicate presentations and light lines, such as drop-shotting and finesse jigging.

Fly-Casting

Fly-fishing, celebrated for its artistic expression and precision, involves using artificial flies to mimic natural prey and attract fish. Revered by passionate anglers, it requires both skill and finesse to excel. At its essence, fly fishing relies on specialized equipment designed specifically for the technique. Fly reels primarily store the line and feature a simple drag system crucial for controlling tension during fish fights. Complementing this, fly rods are known for their length and flexibility, allowing anglers to execute accurate and delicate presentations of their flies to potential catches.

In terms of technique, fly casting is fundamental to this method. It requires employing a specialized casting technique to accurately propel the fly line forward and deliver the artificial fly to the desired target area. This intricate process demands precise timing and coordination between the angler's movements and the rod's action. Crucial to the success of fly-fishing is the art of fly presentation. Anglers must skilfully replicate the behaviour of natural prey, selecting appropriate fly patterns and presentation techniques tailored to the target species and prevailing environmental conditions. Additionally, effective line management is essential, as adept manipulation of the fly line is crucial for achieving casting accuracy and a lifelike presentation. From line mending to controlling its drift and minimizing drag, mastering these aspects ensures a natural presentation that entices fish to bite.

Regulations and Management of Sport Fisheries

Sport fishing regulations play a crucial role in safeguarding aquatic ecosystems, ensuring the sustainable use of fishery resources and maintaining the natural equilibrium of our water bodies. However, in countries like India, there is a notable lack of information and scientific comprehension regarding sport fishing, primarily due to its localized and species-specific nature. While certain Indian states adhere to governmental regulations, there's limited awareness about the importance of conserving threatened and endangered fish species through sport fishing activities. Additionally, promoting this activity could help alleviate unemployment by creating income opportunities and fostering ecotourism. Government intervention is vital for the sustainable development and management of sport fisheries in India (Mandal *et al.*, 2018).

Fishing, both as a general activity and as a sport, has a long history in India, dating back to ancient times as referenced in the Ramayana and Mahabharata periods. In recreational fishing, the choice of equipment and the selection of target species based on their fighting abilities are essential elements of the activity (IGFA, 2016). Anglers engage in a contest with the fish until it's either captured or lost. Early regulatory measures aimed at fish conservation trace back to around 300 B.C. during the Maurya Dynasty, when fishing, including angling, was prohibited from mid-June to mid-December to safeguard breeding and protect juvenile fish. Similarly, during the reign of King Asoka (269–232 BC), fishing was restricted during July and November due to fish breeding seasons (Mandal *et al.*, 2018)

In recent times, various regulatory bodies, like the International Game Fish Association (IGFA), have established rules and guidelines governing recreational fishing, including recent expansions to include lagoon, estuarine, and freshwater species as game fish (IGFA, 2016). These regulations prioritize minimizing stress on fish and reducing catch mortality through practices such as catch-and-release. Additionally, sustainable tourism initiatives should promote biodiversity conservation and poverty alleviation by offering alternative livelihoods and increasing local income (Everard and Kataria, 2011). While ecotourism can supplement household earnings, reliance solely on this sector poses risks due to its susceptibility to global financial fluctuations (Bohensky *et al.*, 2011). Encouraging a variety of livelihood strategies can help households cope with market uncertainties. Recreational fishing programs should also encourage alternative livelihood options to mitigate risks and prevent overexploitation of fish stocks. Introducing new fishing technologies and practices used in sport fishing could benefit local fishers targeting other species, increasing fishing success without excessive pressure on fish stocks (Mandal *et al.*, 2018

Challenges and Opportunities for Sustainable Sport Fishing

Although recreational fisheries in India hold significant potential benefits, numerous issues require immediate attention, as they could hinder the sustainable development of this sector. Some of the major issues have been discussed here.

1) Lack of Information on the Taxonomy of Sport Fish

Freshwater fish in India, including sport fishes, are poorly studied, with limited information on their biology, ecology, and population status (Dahanukar *et al.*, 2011). This includes popular species like the mahseer, which have significant knowledge gaps in taxonomy and natural history despite being documented since the 12th century. The exact number and distribution of mahseer species remain uncertain, with some potentially being species complexes requiring formal taxonomic recognition (Pinder and Raghavan, 2013). Although some studies exist on mahseers, ambiguities in taxonomy and distribution limit their usefulness for conservation planning (Gupta *et al.*, 2015). For species, such as the chocolate mahseer and the goonch catfish, very few biological studies have been conducted, and their complex taxonomy further complicates understanding their distribution. Research on these species is challenging due to their remote habitats, seasonal accessibility issues, and the difficulty of obtaining research permits in protected areas. Recreational fisheries could support research by providing valuable biological data, as demonstrated by a recent study using angler catch data for mahseer conservation in the Cauvery (Pinder *et al.*, 2015).

2) Lack of Understanding of Biotic Responses to Capture and Release

There are no studies on post-capture mortality for mahseer or other angler-targeted species in India. It is speculated, without scientific proof that many may die from exhaustion, injuries, and infections (Dinesh *et al.*, 2010). Fishing gear influences mortality rates (Rocklin *et al.*, 2014); natural baits lead to higher mortality due to deeper hooking (Wilde *et al.*, 2000), whereas circle hooks and barbless hooks lower mortality by encouraging shallow hooking and reducing handling time. The lack of data on the impact of catch-and-release (C&R) practices on popular sports fish in India makes it difficult to evaluate the sustainability of these activities. Research is necessary on post-release mortality rates and factors affecting mortality to create best practices for responsible and sustainable C&R (Cooke and Suski, 2005).

3) Need for Development and Dissemination of Best Practices for Sustainable Angling Promotion

At present, India lacks official guidelines and monitoring for recreational fisheries, placing the responsibility on angling associations, some of which endorse best management practices. For instance, the 2014 "All India Fresh Water Angling Competition" provided participants with guidelines, including mandatory catch-and-release (C&R) rules. However, not all associations adhere to these standards. A quick survey by an author uncovered harmful practices such as using wall nails as hooks, excessive air exposure for fish, and visible wounds on fish intended for release. To address these issues, increased awareness through angler education and enforcement by a statutory recreational angling body is necessary (Gupta *et al.*, 2015).

4) Unregulated Stocking and Introductions

The development of recreational fisheries in India has been bolstered by the introduction and stocking of both exotic and captive-bred native species. During the British Raj, upland lakes and river reaches were stocked with exotic salmonids like brown trout and rainbow trout to promote recreational fishing. Only brown trout established self-sustaining populations, however, affected the native snow trout population (Sehgal, 1999a). Similar stocking efforts took place in the Western Ghats, where trout hatcheries were established, and recreational trout fishing continues with the support of local angling associations (Sehgal, 1999b). The biological and socio-economic impacts of these practices, especially in biodiversity-rich areas, remain poorly understood. There is a need to evaluate anglers' preferences and awareness regarding native and non-native species and to assess support for stocking to enhance recreational fishing (Nguyen *et al.*, 2013). Understanding stakeholders' attitudes can help guide management and conservation efforts (Gupta *et al.*, 2015).

5) Poorly Defined Governance Structures

The growth of India's recreational fisheries sector has been impeded by its multi-jurisdictional governance structure, which lacks a centralized body to streamline decision-making at both national and state levels. This fragmentation has created difficulties in the oversight, management, and regulation of sustainable fisheries. While many angling associations operate with permits, unlicensed groups remain common, and there is a disconnection between fisheries management associations and government agencies (Gupta *et al.*, 2015). For example, in the Himalayan region, the Forest Department issues fishing permits but cannot effectively patrol fishing areas. Additionally, disputes over revenue distribution from fishing permits among the Forest Department, angling associations, and local communities have led to suggestions that the Uttarakhand Fisheries Department should manage recreational angling (Gupta *et al.*, 2014). Recreational angling tourism provides social and economic benefits to local communities, but transparency issues in profit-sharing have been raised. A proposed conservation tax on recreational anglers to support local communities has been well-received, but restricting villagers from catching food fish in protected areas has caused concerns.

6) Poor Stakeholder Engagement

Many anglers in India have highlighted the lack of government support for recreational fisheries and stressed the importance of establishing angling conservation units within village communities to ensure local stakeholders' benefit. They also expressed concerns about law enforcement, noting that illegal fishing often goes unpunished due to the absence of formal protection strategies for crucial fish habitats. Additionally, there is a pressing need for scientific studies to understand the impact of recreational angling on freshwater biodiversity to raise public awareness about these ecosystems (Gupta *et al.*, 2015). To address the ongoing conflict between anglers and village members over access rights, a common platform for dialogue is necessary. One proposed solution is to set up freshwater fish-safe zones monitored by local communities, which could provide both recreational and sustenance opportunities (Gupta *et al.*, 2014). Legislative support at the central or state level for recreational angling could provide structure and highlight its benefits, but this must be linked with freshwater conservation efforts to control illegal fishing practices and the introduction of exotic fish species.

7) Conflict between Recreational Fisheries and Other Activities

Anglers have observed destructive fishing methods like the use of explosives, illegal nets, poisons, and electricity near their fishing spots. They pointed out various threats to freshwater ecosystems, including overfishing, illegal fishing practices, water pollution, insufficient administrative support, poor freshwater management, habitat clearing, hydro-electric projects, and the introduction of exotic fish species (Raghavan *et al.*, 2012)

8) Lack of Representative Data from the Recreational Fishery Sector

A major obstacle in managing recreational fisheries in India is the absence of representative data, stemming from the challenges of sampling people across different regions. Traditional methods like telephone-diary surveys are difficult to execute, making it necessary to explore alternatives such as face-to-face interviews or angler diaries. Techniques currently being tested in Australia, as well as various sampling methods used in health sciences, may be relevant. It is essential to involve government, fishing organizations, and communities in funding these surveys to achieve a representative sample of Indian recreational fishers (Gupta *et al.*, 2015).

Opportunities

In the recent times, recreational fishing particularly angling has been recommended by medical authorities as a healthy pursuit and emerged as an outdoor recreation (Mandal *et al.*, 2018). Both socio and economic benefits have been realised from the recreational fishing, which may be substantial for the developing countries (Pinder and Raghavan, 2013). Study revealed that recreational angling in India is mainly a male-dominated activity due to societal norms. Angling associations could promote female participation by providing opportunities for women and making the sport more female-friendly, which could also enhance local economies through tourism (Gupta *et al.*, 2015). Managing freshwater fisheries faces governance challenges, particularly concerning jurisdiction over water bodies within protected areas (PAs). Forest managers often claim control over entire water bodies, causing conflicts with local stakeholders who advocate for the involvement of state fisheries departments. Cooperative management and consensus between forest departments and local communities are crucial (Gupta *et al.*, 2014). Stocking and ranching practices must be scientifically evaluated to prevent a reduction in genetic diversity. According to IUCN guidelines, reintroductions should benefit both the species and their ecosystems, rather than merely increasing catch sizes for anglers. Recreational fisheries management in India must develop to provide long-term ecological, social, and economic benefits while supporting wider freshwater conservation efforts (Rogers, 2013).

Conclusion

Recreational fisheries in India, with a specific focus on the Kashmir Valley, reflect a diverse and vibrant sector that offers both ecological benefits and socio-economic opportunities. Sport fishing, characterized by its pursuit of leisure rather than sustenance, has evolved significantly in India, driven by the allure of challenging game fish like mahseer and trout. The Kashmir Valley boasts a rich tapestry of rivers, streams, and high-altitude lakes that attract enthusiasts from around the world. Despite its allure, recreational fishing in India faces several challenges. These include the lack of comprehensive data on fish taxonomy and biotic responses to catch-and-release practices, as well as governance issues stemming from fragmented regulatory frameworks and insufficient stakeholder engagement. Addressing these challenges is crucial for sustainable management and conservation of India's aquatic biodiversity. However, there are promising opportunities for the development of recreational fisheries in India. These include promoting eco-tourism, enhancing community involvement in conservation efforts, and fostering gender-inclusive participation in angling activities. By implementing sound management practices, fostering scientific research, and improving regulatory frameworks, India can harness the full potential of recreational fisheries while safeguarding its valuable aquatic resources for future generations.

References

Baruah, D., & Sarma, D. (2018). Mahseer in recreational fisheries and ecotourism in India. NACA Newsl, 22(2), 1-10.

Bohensky, E. L., Butler, J. R., & Mitchell, D. (2011). Scenarios for knowledge integration: exploring ecotourism futures in Milne Bay, Papua New Guinea. Journal of Marine Sciences, 2011(1), 504651.

Cooke, S. J., & Suski, C. D. (2005). Do we need species-specific guidelines for catch-and-release recreational angling to effectively conserve diverse fishery resources?. Biodiversity & Conservation, 14, 1195-1209.

Dinesh, K., Nandeesha, M. C., Nautiyal, P., & Aiyappa, P. (2010). Mahseers in India: A review with focus on conservation and management. The Indian journal of animal sciences, 80(4).

Everard, M., & Kataria, G. (2011). Recreational angling markets to advance the conservation of a reach of the Western Ramganga River, India. Aquatic Conservation: Marine and Freshwater Ecosystems, 21(1), 101-108.

FAO. (2012). Recreational fisheries: FAO technical guidelines for responsible fisheries.

Froese, R. and D. Pauly. Editors. 2024. FishBase. World Wide Web electronic publication. www.fishbase.org, (02/2024)

Gupta, N., Bower, S. D., Raghavan, R., Danylchuk, A. J., & Cooke, S. J. (2015). Status of recreational fisheries in India: development, issues, and opportunities. Reviews in Fisheries Science & Aquaculture, 23(3), 291-301.

Gupta, N., Sivakumar, K., Mathur, V. B., & Chadwick, M. A. (2014). The 'tiger of Indian rivers': Stakeholders' perspectives on the golden mahseer as a flagship fish species. Area, 46(4), 389-397.

Henderson, F. R. (2000). Let's go fishing: a fish and fishing project manual.

Hora, S. L. (1951). Knowledge of the ancient hindus concerning fish and fisheries of India. J. Asiat. Soc. Lett, 17, 145-169.

International Game Fish Association (IGFA), 2016. IFGA Fish Database. Florida (USA), (http://igfa.org/fish/fishdatabase.aspx)

Mahanta, P. C., Sarma, D., & Joshi, K. D. Sport Fishing and Livelihood in Upland Himalayan Region.

Mandal, A., Das, S. K., Biswas, B., & Kairnar, S. O. (2018). Future scope of sport fisheries development in India: a review. Indian Journal of Animal Health, 57(1), 1-16.

Molur, S., Smith, K. G., Daniel, B. A., & Darwall, W. R. T. (2011). The status and distribution of freshwater biodiversity in the Western Ghats, India. Cambridge, UK and Gland, Switzerland: IUCN, and Coimbatore, India: Zoo Outreach Organisation.

Nautiyal, P. (2014). Review of the art and science of Indian mahseer (game fish) from nineteenth to twentieth century: road to extinction or conservation?. Proceedings of the National Academy of Sciences, India Section B: Biological Sciences, 84, 215-236.

Nguyen, V. M., Rudd, M. A., Hinch, S. G., & Cooke, S. J. (2013). Recreational anglers' attitudes, beliefs, and behaviors related to catch-and-release practices of Pacific salmon in British Columbia. Journal of environmental management, 128, 852-865.

Pinder, A. C., & Raghavan, R. (2013). Conserving the endangered mahseers (Tor spp.) of India: the positive role of recreational fisheries. Current Science, 104(11), 1472-1475.

Pinder, A. C., Raghavan, R., & Britton, J. R. (2015). Eff icacy of angler catch data as a population and conservation monitoring tool for the f lagship Mahseer f ishes (Tor spp.) of Southern India. Aquatic Conservation: Marine and Freshwater Ecosystems, 25(6), 829-838.

Raghavan, R., Ali, A., Dahanukar, N., & Rosser, A. (2011). Is the Deccan Mahseer, Tor khudree (Sykes, 1839)(Pisces: Cyprinidae) fishery in the Western Ghats Hotspot sustainable? A participatory approach to stock assessment. Fisheries Research, 110(1), 29-38.

Rocklin, D., Levrel, H., Drogou, M., Herfaut, J., & Veron, G. (2014). Combining telephone surveys and fishing catches self-report: The French sea bass recreational fishery assessment. PloS one, 9(1), e87271.

Rogers, A. A. (2013). Social welfare and marine reserves: Is willingness to pay for conservation dependent on management process? A discrete choice experiment of the Ningaloo Marine Park in Australia. Canadian Journal of Agricultural Economics/Revue canadienne d'agroeconomie, 61(2), 217-238.

Sehgal, K. L. (1999a). Coldwater fish and fisheries in the Indian Himalayas: rivers and streams. Fish and fisheries at higher altitudes: Asia. Food and Agriculture Organization of the United Nations Technical Paper, 385, 41-63.

Sehgal, K. L. (1999b). Coldwater fish and fisheries in the Western Ghats, India. Fish and fisheries at higher altitudes: Asia, 385, 103.

Wilde, G. R., Muoneke, M. I., Bettoli, P. W., Nelson, K. L., & Hysmith, B. T. (2000). Bait and temperature effects on striped bass hooking mortality in freshwater. North American Journal of Fisheries Management, 20(3), 810-815.

14

Fish Health and Disease Management Practices in Kashmir Valley

Kafeela Mukhtar Qureshi[1*], Feroz Ahmad Shah[1], Irtifa Fayaz[1], Shabir Ahmad Dar[1], and Zaheer Abbas[2]

[1]Division of Aquatic Animal Health Management, Faculty of Fisheries, SKUAST-K Rangil, Ganderbal, J&K, India
[2]Division of Fisheries Resource Management, Faculty of Fisheries, SKUAST-K Rangil, Ganderbal, J&K, India

Abstract

Disease outbreaks and problems with fish health pose serious hurdles to the aquaculture industry in Kashmir Valley. This book chapter explores the approaches to managing fish health and controlling disease, emphasizing common infections, diagnosis methods, and treatment strategies. Both conventional and modern methods such as the application of chemotherapeutics, antibiotics, and environmental control are assessed critically. Concerns concerning resistance and environmental impact are raised by the rising use of antibiotics in aquaculture, which has resulted in a rise in illness incidence. The study emphasizes the necessity of integrated disease management plans that safeguard fish populations and the surrounding ecosystem by combining efficient treatments with long-term methods. To facilitate the application of these strategies, better infrastructure and increased training for regional aqua culturists are advised. This abstract examines the Kashmir Valley's current methods for managing fish health, emphasizing the need for more technical know-how and better infrastructure to support the expansion and sustainability of the area's fishing sector.

Keywords: *Disease management, Fish health, Kashmir Valley, Chemotherapeutics, Antibiotics*

Introduction

Fish raised in cold water fisheries are species that flourish in temperatures usually below 20°C. To increase fish production and provide jobs for the local population, cold water fisheries have been established throughout India. One important region in India where cold-water fisheries have been effectively established is Himachal Pradesh. There are plenty of cold-water rivers and streams in this area that are perfect for fish farming. To encourage cold water fisheries, the state government has launched several programmes, including financing for the construction of fish farms and the purchase of supplies and equipment. The most often farmed fish species in Himachal Pradesh's cold water fisheries are brown trout, rainbow trout, and mahseer (Bhat *et al.*, 2023). India is well known for its cold-water fish, which are highly prized for their flavour. In addition to Himachal Pradesh, other states that have embraced cold-water fisheries include Jammu and Kashmir, Uttarakhand, Sikkim, and Arunachal Pradesh. However, obstacles to their growth include restricted market access, a lack of technological know-how, and inadequate infrastructure. The many cold-water fisheries in northern and northeastern India demonstrate the region's enormous potential despite these obstacles (Pandey *et al.*, 2021).

These fisheries support a variety of fish species that have evolved to live in colder waters. With an approximate length of 2,500 km, the Himalayas stretch from Nanga Parbat (8,126 m) in the west to Namcha Barwa (7,756 m) in the east, encompassing an area of 594,400 km^2. Its western boundary is the high plateau of Tibet; its northern boundary is the Karakoram Mountains; the range's southern and northern limits are separated by 200–400 km. The Siwaliks, the lesser and Greater Himalayas, and the Trans-Himalaya run parallel to each other from south to north and are identified by their physiographic features and geological histories. With a combined catchment area of around 160,000 km^2, the Indus and Brahmaputra are the longest of the nineteen major rivers that meander through the Himalayas. When these rivers emerge from the mountains; many of them pass through wide valleys. According to Shinn *et al.* (2015), the distribution of fish species in Himalayan streams is influenced by variables such as flow velocity, substratum type, water temperature, and the availability of food. The intensification of aquaculture systems has resulted in the appearance of new pathogens and a greater dependence on antibiotics and other supplements, which threaten the health of fish, humans, and the environment (Watts *et al.*, 2017). Aquaculture faces a great deal of difficulty due to diseases, which result in large production losses that surpass six billion USD annually (Bhat *et al.*, 2023). According to estimates, 10–50% of aquaculture production expenses are related to disease management (Sahoo *et al.*, 2017). This book chapter provides a brief overview of some prevalent parasite, viral, and bacterial diseases that have been linked to temperate and cold-water aquaculture, along with information on potential remedies and preventative measures. Commonly disease found in fish are presented in (Table 1, 2 and 3).

Table 1: Some of the commonly found bacterial diseases in the fishes of Kashmir (Modified from Bhat *et al.*, 2023)

Sr. No.	Disease	Aetiological agent	Susceptible age group	Signs
1	Motile Aeromonas septicaemia (MAS)	*Aeromonas hydrophila*	All age groups	Haemorrhagic septicaemia
2	Rainbow trout fry syndrome (RTFS)	*Flavobacterium psychrophilum*	Fry	Loss of equilibrium, erratic swimming, lethargy, heightened excitability, hemorrhaging, and mortality
3	Columnaris disease/ Saddle-back disease	*Flavobacterium columnare*	All age groups	Lesions at the base with a 'saddleback' appearance, resembling cotton wool. around mouth or mouth fungus
4	Cold-water disease (CWD)	*Flavobacterium psychrophilum*	All age groups	Loss of fin tip integrity, accompanied by darkening of the skin color in the peduncle region.
5	Bacterial gill disease	*Flavobacterium* sp.	Mostly fry and fingerling	Lethargy and gasping for air, with fish aggregating near the inlet of the tank.
6	Enteric Red Mouth Disease	*Yersenia ruckeri*	All age groups	Lethargy, skin darkening, and congestion around the mouth and operculum
7	Mycobacteriosis (Fish Tuberclosis)	*Mycobacterium*	All age groups	Zoonotic poteintial,

Table 2: Some of the commonly found fungal disease in the fishes of Kashmir

Sr. No.	Disease	Aetiological agent	Susceptible age group	Signs
1	White spot disease	*Ichthyophthirius multifiliis*	Fry and fingerlings	White pinhead-sized patches appear on the epidermis and gills
2	Trichodiniasis	*Trichodina, Trichodinella and Tripartiella*	Common in young fish	Tiny red dots appearing on the gills.

3	Gyrodactylosis	*Gyrodactylus salaris*	All age groups	Excess mucus secretion, gill filament breakage, and hemorrhagic lesions in the gills.
4	Black spot disease	*Diplostomum sp*	Fry and fingerling commonly affected	Each fish exhibits a unique pattern of spots, varying in size from 1 to 2 mm, depending on the species.
5	Eye disease of rainbow trout	Digenetic trematode (metacercaria)	Adult trout	Cataracts, lens displacement and degeneration, cortical fibre liquefaction and eventually blindness
6	Argulosis	Argulus sp	All age groups	Confined reddish spots on the outside of the body
7	Whirling diseases	*Myxobolus cerebralis*	Small fry and fingerlings	Inconsistent swimming due to imbalance and loss of equilibrium

Table 3: Some of the majorly occurring fungal diseases found in the fishes of Kashmir (DCFR)

Sr. No.	Disease	Aetiological agent	Susceptible age group	Signs
1	Saprolegneiosis	*Saprolegnia sp*	All life stages including eggs	White mat appearance, lethargy, listless and less responsive to external stimuli
7	Bhranchiomycosis	*Branchiomyces sanguinis, B. demigrans*	Small fry fingerlings and adult fish	Discoloration, disarray, partial gill loss, and ischemia.

Control Measures for Cold-Water Fish Diseases

To prevent the emergence of bacteria that are resistant to antibiotics, a large number of researchers are currently looking for appropriate and sustainable substitutes for treating microbial illnesses in aquaculture. Probiotics, vaccinations, phytotherapy (Bhat *et al.*, 2021), antimicrobial peptides (Chaturvedi *et al.*, 2020), and others have recently been suggested as substitutes. Here, we provide a quick overview of the substances and medications now employed to manage fish illnesses (Sahoo *et al.*, 2017; Bhat et al., 2023) (Table 4). Medication can be given by feed or by adding it to the water, depending on the disease and the population of fish impacted. The following techniques can be applied to efficiently deliver the necessary therapeutic dose for illness control (Bhat *et al.*, 2023).

Disease Prevention

To minimise production losses, disease outbreaks in fish and shellfish farming must be mitigated through the development and application of efficient diagnostic and control measures. Managing culture units to reduce environmental factors that predispose animals to sickness is a crucial method in disease management programmes. In order to do this, it is necessary to maintain suitable stocking densities, avoid the introduction of pathogens into hatcheries or culture systems, ensure optimal water quality, reduce stress, and provide farmed animals with sufficient nourishment (Das *et al.*, 2014).

Pond Management: Effective pond management is essential for enhancing environmental conditions, preventing disease, and increasing fish yields. There are two primary components to pond management: pond maintenance and pond disinfection (Das *et al.*, 2014).

Fish Rearing Supervision: Every day pond maintenance chores including feeding, stocking, preventing disease and other activities should be supervised by a responsible person. Fish productivity and illness management both benefit from the use of a consistent feeding schedule. It's crucial to keep an eye on changes in water quality. Additionally, depending on local conditions, fish seeds can be disinfected either before stocking or during transfer, usually with a bucket or jar (Raja *et al.*, 2015).

Pond treatment: To prevent illnesses, chemicals are used all around ponds to treat them. Thorough cleaning and disinfection of ponds with a 1 ppm potassium permanganate solution is necessary before stocking fish seed. To improve the quality of deteriorated pond water and prevent illnesses, quicklime should be dissolved in water, well agitated, and distributed evenly around the pond. But as the freshwater farming sector grows quickly and the movement of fry and fingerlings across provinces increases, regional illnesses have a tendency to spread. In these circumstances, it is definitely forbidden to transfer sick fish and the quarantine should be adhered to (Bhat *et al.*, 2023).

Diagnosis: Correct identification of a condition is necessary for efficient therapy, hence handling must be done with extreme caution. Fish should be kept moist, and they should be living or recently deceased. It is best to preserve dissected organs as much as possible. To avoid germs from spreading between organs, post-mortem examination instruments need to be properly cleansed and sanitised. For microscopic analysis, observe skin samples in distilled water and internal organs in 0.85% normal saline. If difficulties are encountered after diagnosis, distinguish between primary and secondary disorders and treat them independently or in tandem as necessary (Vijayan and Salin, 2012).

Methods of Treatments

Oral Route: Medication administered through fish feed must be handled carefully to prevent it from leaching into the water without reaching its intended target. Chemotherapeutics can be mixed into the feed in precise amounts and fed to the fish. However, this method is most effective when applied early in the disease process, as severely affected fish may not eat. Antibiotics like sulfa drugs, oxytetracycline, and nitrofurans are frequently incorporated into feed to treat bacterial infections. The feed can be suspended in perforated containers for administration (Bhat *et al.*, 2023).

Dip Treatment: A more concentrated application of chemicals or disinfectants is employed in a process designed to treat small-sized fish under controlled conditions. The exposure duration typically lasts less than 1–2 minutes (Nayak and Mogalekar, 2022). The primary benefit of this approach is its ability to treat a large population with minimal medication doses.

Flush Treatment: At the inlet or entry point where incoming water flows into the system, a chemical is introduced and allowed to circulate throughout the tank. For raceway units with a high flow rate, this method is effective in completely flushing out the drug within a specified time frame (Noga, 2010).

Bath Treatment: This method is applied in small experimental tanks under close supervision. After introducing chemicals or medications, they are left for a specified duration. If fish show signs of stress, either freshwater is supplied to the tank, or the fish are transferred and released into freshwater (Bhat *et al.*, 2023).

Table 4: Some of the most commonly used disinfectants, drugs and chemicals used in the fisheries sector of Kashmir (Modified from Sahoo *et al.*, 2017 and Bhat *et al.*, 2023)

Name	Quantity required (Dosage)	Type of treatment
Potassium Permanganate	20–30 kg/ha-m 25–50 g in 50 L	Pond treatment at 2–3 ppm Bath treatment at 500–1000 ppm for 1–2 min
Oxytetracycline	70 g for 1000 kg	Administer feed at a dosage of 70 mg/kg of body weight for 10 days.
Chloramphenicol	55 g for 1000 kg	Mix with feed at 50 mg/kg body weight for 10 days
Chlortetracycline	55 g for 1000 kg	Administer feed at a dosage of 55 mg/kg of body weight for 10 days.
Erythromycin thiocyanate	45 mg/lb	Administer feed at a dosage of 100 mg/kg of body weight per day for 21 days.
Sulfadimethoxine/ ormetoprim	23mg Sulfadimethoxine/ ormetoprim	Feed at a dosage of 50 mg/kg of body weight per day for 5 days.

Malathion 50%	5 L/ha-m	Pond treatment at a concentration of 0.25 parts per million (ppm) administered in three separate doses.
Butox	70–75 mL/ha-m	Pond treatment
Florfenicol	10 grams for 1 tonne of fish	Administer feed at 10 mg/kg (4.5 mg/lb) of body weight per day for 10 days to treat furunculosis or rainbow trout fry syndrome. For treating pasteurellosis, provide feed at a dosage of 6–25 mg florfenicol/kg (2.7–11 mg/lb) of body weight per day.
Enrofloxacin	Feed at 4.5 mg/lb. Inject 2.2–4.5 mg/lb	Administer feed at a dosage of 10 mg/kg of body weight per day for 10 days. For preoperative infection prevention, inject 5–10 mg per body weight intramuscularly (IM).
Bronopol (Pyceze)	20 mg bronopol/ l = 0.15 mL Pyceze™/ gallon 50 mg bronopol/ L = 0.38 mL	Add 1 mL of Pyceze per 25 L of water and treat for 30 minutes. Alternatively, add 0.1 mL of Pyceze™ per liter and treat for 30 minutes daily, starting 24 hours after fertilization, to treat water mold infections on eggs.
Sodium chloride	2–3 kg in 100 L	Bath treatment at 2–3% for 1–2 min
Formalin	150–250 L/ha-m 5 mL in 20 L	Pond treatment at 15–25 ppm Dip treatment at 250 ppm for 10 min
Acriflavin	10 g in 20 L	Bath treatment at 500 ppm for 20 min

Conclusion

In conclusion, effective fish health and disease management practices are crucial for sustaining aquaculture in the Kashmir Valley. Through this study, it becomes evident that proactive measures such as maintaining water quality, implementing biosecurity protocols, and conducting regular health monitoring are essential for preventing disease outbreaks and ensuring the overall well-being of fish populations. Additionally, promoting awareness among fish farmers about early detection methods and appropriate treatment options can significantly enhance the resilience of the aquaculture industry against potential health challenges. By adopting these practices, stakeholders can not only safeguard fish health but also contribute to the sustainable growth of aquaculture as a vital component of the region's economy and food security. Disease surveillance and monitoring systems should be implemented to closely monitor disease outbreaks and the emergence of new pathogens. A thorough grasp of disease prevalence, local disease prevention and control technologies, the formulation of cost-effective biosecurity programs, and the implementation of best management practices (BMPs) and husbandry measures at the farm level are essential components that require significant focus to ensure sustainable fisheries production in the region.

References

Ananda Raja, R., and Jithendran, K. P. (2015). Aquaculture disease diagnosis and health management. In Advances in marine and brackishwater aquaculture (pp. 247-255). New Delhi: Springer India.

Bhat, R. A. H., Mallik, S. K., Tandel, R. S., & Shahi, N. (2023). An overview of cold-water fish diseases and their control measures. Fisheries and Aquaculture of the Temperate Himalayas, 255-283.

Chaturvedi, P., Bhat, R. A. H., & Pande, A. (2020). Antimicrobial peptides of fish: Innocuous alternatives to antibiotics. Reviews in aquaculture, 12(1), 85-106.

Das, B. K., & Mishra, S. S. (2014). Diseases in Freshwater aquaculture. Training Manual on Model training course on Preventive health management practices in freshwater aquaculture. ICAR-Central Institute of Freshwater aquaculture, Bhubaneswar, Odisha, India.

Mishra, S. S., Rakesh, D., Dhiman, M., Choudhary, P., Debbarma, J., Sahoo, S. N., & Mishra, C. K. (2017). Present status of fish disease management in freshwater aquaculture in India: state-of-the-art-review. Journal of Aquaculture & Fisheries, 1(003), 14.

Nayak, S. K., & Mogalekar, H. S. (2022). Common freshwater Fish diseases and their control measures. Vigyan Varta An International E-Magazine for Science Enthusiasts.

Noga, E. J. (2010). Fish disease: diagnosis and treatment. John Wiley & Sons.

Pandey V, Bhat RAH, Chandra S, Tandel RS, Dubey MK, Sharma P, Gehlot B, Dash P, Joshi R (2021) Clinical signs, lethal dose and histopathological lesions in grass carp, Ctenopharyngodon idella experimentally infected with Edwardsiella tarda. Microb Pathog 161:105292

Shinn, A., Pratoomyot, J., Bron, J., Paladini, G., Brooker, E., & Brooker, A. (2015). Economic impacts of aquatic parasites on global finfish production. Global Aquaculture Advocate, 2015, 58-61.

Vijayan, K. K., & Sanil, N. K. (2012). Health Management Strategies for profitable and sustainable Aquaculture, with special reference to diagnostics. Lead papers on strategies for Aquaculture Development. ICAR-Central Institute of Freshwater Aquaculture, Bhubaneswar, India.

Watts, J. E., Schreier, H. J., Lanska, L., & Hale, M. S. (2017). The rising tide of antimicrobial resistance in aquaculture: sources, sinks and solutions. Marine drugs, 15(6), 158.

15

Fisheries Sector in Kashmir Valley The Threats, Opportunities and Challenges

Asifa Wali* and Tasaduq H. Shah

Sher-e-Kashmir University of Agricultural Sciences and Technology Kashmir, Faculty of Fisheries, Rangil-190006, Ganderbal, J&K

Abstract

The Union Territory of Jammu and Kashmir, endowed with a diverse range of water resources spread across 0.40 lakh hectares, presents significant potential for cold-water, warm-water, reservoir, and sport fishing due to its unique topography and climate. Despite contributing to three-quarters of the nation's cold-water fisheries, J&K accounts for only 1% of India's aquaculture output. The region faces challenges such as limited infrastructure, environmental degradation, and the impacts of tourism and urbanization, which threaten the sustainability of its fisheries. Native species, particularly in the Schizothorax genera, are at risk due to pollution, eutrophication, and the invasion of alien species. This chapter explores the effects of human activities on cold-water fisheries and emphasizes the critical need for sustainable management practices to safeguard these essential ecosystems. Emphasizing the importance of local community involvement, the study advocates for the implementation of effective sewage treatment, organic farming practices, and strict regulations to conserve water bodies and their biodiversity.

Keywords: *Jammu and Kashmir, Cold-water fisheries, Environmental degradation, Sustainable management, Aquatic ecosystems*

Introduction

The Union Territory of Jammu and Kashmir is blessed with various water resources dispersed over an area of 0.40 lakh hectares due to its distinct terrain and three distinct climatic conditions. There is enormous potential for cold-water, warm-water, reservoir, and sport fishing due to the existence of cold-water streams, rivers, lakes, high-altitude lakes, springs, and sars, among other features (DoF-J&K, 2022). Three-quarters of the cold-water fisheries are derived from J&K. Still, the UT makes for no more than 1% of the nation's aquaculture output. According to DoF (2020), J&K had 0.21 lakh tonnes of inland fish output in 2019–20. In Kashmir, just 15% of the 70% of people employed in agriculture are fishermen (Qureshi, 2013). Limited infrastructure and connectivity in the highlands are substantial obstacles to the sector's expansion. However, the fishing industry still plays a significant role in the existence of the inhabitants of Kashmir, providing subsistence fishing as a means of nutrition and operating as an affordable and dependable supply of protein in the landlocked area (CMFRI, 2015). Approximately 70% of the entire output comes through the lakes of Dal and Wular, which account for the bulk of the harvest (Qureshi, 2013).

The characteristics of cold-water assets, including creeks and upland streams, contain exceptional transparency, high dissolved oxygen (DO), bedrock-boulder-sand substrata, and high flow rates. These natural settings consist of gushing waters, cascades, torrents, and deep pools. The slopes of the Himalayas are home to 272 fish species (Sarma *et al.*, 2018). Elevated heights result in reduced human influence and nutrient-poor water bodies due to catchment features. The glacier lakes are ultra-oligotrophic at relatively high elevations (Gopal, 2012). Fish in these waters develop slowly and only to tiny sizes because of the limited biological production. Large-scale commercial fisheries are therefore not supported by these cold-water resources. Likewise in other areas, human footprints have had a noticeable effect on Kashmir's flora, wildlife, and water sources. Challenges include things like contamination, eutrophication, and the inundation of alien species. Massive sedimentation has been made worse by tourism and growing urbanisation, which has reduced fisheries. Particularly impacted are native fish species, notably those in the *Schizothorax* genera (Qureshi *et al.*, 2013). The future value of resourcefulness is diminished by this deterioration, which affects regional populations' ability to make a living and provide adequate nutrition. At this stage, the lakes in Kashmir are frequently contaminated, weed-filled, vulnerable to harmful algal blooms, bordered by marshlands (Tahseen *et al.*, 2018). The concern must be addressed, with special attention to the larger context of Kashmir. Under several topics, this section examines the primary human interactions and how they affect the cold-water fishing industry and management.

1. Exotic Species Addition

Native throughout Kashmir, *Schizothorax* genus came there through streams in Central Asian countries (Sehgal, 1999). These were confined to the desolate area bounded by the Sulaiman lineup, the northwestern Himalayas, the inner and southern sides of the Hindu Kush, and Karakorum (Raina & Peter, 1999). The aforementioned species have become an essential component of the fishery in Jammu and Kashmir because they can survive in clear, icy waters. Eventually, in 1959, common carp were introduced into Kashmiri lakes, changing their position (Sehgal, 1989). Schizothorax and common carp have comparable environments and reserves of food, which increases rivalry. The common carp surpasses Schizothorax because to its faster rate of reproduction and tolerance to mating in limited waterways (Singh and Lakra, 2008). After common carp were introduced, Dal Lake's fishery expanded dramatically, however, its indigenous species' reproduction environments suffered as a result of the expansion. Wular Lake's output is similarly affected, however considering the dimensions of the lake, sludge accumulation might have had a greater influence than the addition of foreign species (Qureshi *et al.,* 2013). Furthermore, it is now documented that mirror carp have infiltrated the ecological niches of *Osteobrama belangerii* in Loktak Lake, Manipur; *Schizothorax* within the Kashmir highlands; and Gobindsagar Reservoir, Kumaon Himalayas (Swain, 2017). The dire need for common carp is highlighted by a higher appetite for fish, independent of species, even though fishermen, who are the primary interest groups, favour *Schizothorax* species. Rather than giving fishermen's concerns top priority, the Department of Fisheries (DoF) is concentrating more on generating cash employing the tourism industry.

It was previously documented that brown trout (*Salmo trutta fario*), which were introduced to the Harwan hatchery in Kashmir in 1900, put young *Schizothorax* at risk of extinction by feeding on them (Kausar *et al.*, 2017). According to the NRCWF (2002), brown and rainbow trout account for 40–50% of the state's earnings and are important for fishing and tourism. The eastern brook trout, splake trout and the solitary version of Atlantic salmon from North America are some other salmonid species that have been introduced in Kashmir. Only splake trout has not been able to root out itself among these. Yet, rainbow trout only survives in hatchery facilities, brook trout and landlocked Atlantic salmon are equally constrained, while brown trout have found ways to live in stream ecosystems. As such, the brown trout presents a serious threat to the continued existence of native species. To stop conflict for nutrients and hybridisation with native species, it is imperative to keep such exotic species contained within hatching facilities and out of streams and lakes.

Conservation and Management of *Schizothorax* Species

Reports indicate that *Schizothorax* populations have declined by up to 90% in some areas, with an overall reduction of 50% or more, and this trend is expected to continue, classifying them as a Vulnerable species. Currently, there are no established research, monitoring, education, land/water protection, species management, or invasive species control measures in place (IUCN, 2010). This situation calls for a balanced approach between indigenous species and common carp. Reducing the common carp population while utilizing alternative resources for their culture can ensure a stable income for culturists and maintain the fish supply chain. There is a need to focus on ranching programs and the development of reliable hatchery technology for Schizothorax species. Efforts should continue to produce viable Schizothorax fingerlings that can be successfully released into lakes and rivers (Sehgal, 1999). Current efforts are directed towards *S. niger* and *S. esocinus* (Sarma *et al.*, 2018). Additionally, lake fisheries should be organized with a robust end-to-end supply chain, and habitat conditions must be improved to support the growth and development of Schizothorax populations.

2. Effect of Eutrophication

Eutrophication refers to the process where the levels of organic and inorganic dissolved and particulate matter in water bodies increase significantly, leading to heightened biological productivity but reduced overall water quality. The extent of eutrophication can be assessed using various methods, including the trophic state index (TSI) based on algal biomass, Secchi disk transparency, algal chlorophyll concentration, and total phosphorus levels (Carlson, 1977). The nutrient status of a water body influences aquatic life, including its distribution and reproduction, by affecting the physical, chemical, and biological processes within the environment. However, human activities such as industrialization, urbanization, and pollution (Fakayode, 2005) disrupt this balance, leading to increased biochemical oxygen demand (BOD), chemical oxygen demand (COD), alkalinity, and hardness (Kulkarni, 1997). Eutrophication is a process driven by the system's biochemistry, including growth, decay, and the internal phosphorus cycle, and cannot be completely halted. Any additional anthropogenic nutrient loading exacerbates eutrophication, resulting in nearly irreversible changes to the water body (Lone *et al.*, 2022).

Nitrogen and Its Impact on Water Quality

Nitrogen plays a crucial role in biological growth, supporting the development of plants, fungi, and other living organisms (Mlitan *et al.*, 2015). It exists in water in various forms, including nitrites (NO_2^-), nitrates (NO_3^-), ammonium (NH_4^+), and ammonia (NH_3), which indicate levels of pollution and micronutrients. Natural processes such as nitrification, denitrification, and the mineralization of organic matter contribute to nitrogen levels in freshwater systems. Surface waters often become polluted with nitrate nitrogen due to wastewater discharge from agricultural runoff, industrial discharges, municipal/domestic sources, and animal feedlots. This pollution is a significant factor in the eutrophication of surface waters (Fried, 1991). A study by Bhat et al. (2017) in Dal Lake found that nitrate nitrogen levels were highest in summer, followed by spring, and lowest in autumn. The reduced nitrate concentration in autumn was attributed to lower organic load, decreased decomposition of organic matter, and less agricultural activity during that season. The study also identified the highest nitrate concentrations in areas with significant urban sewage discharge and agricultural runoff, highlighting the impact of human activities on nutrient loading in Dal Lake.

Role of Phosphates in Eutrophication

While nitrates contribute to eutrophication, phosphates are even more critical in triggering the process. Although phosphates are essential for living organisms, they become pollutants at high concentrations. Phosphates exist in several forms, including orthophosphates, condensed phosphates, and organic phosphates, and can be found in solution, in organisms, particulate matter, and detritus (Rapin *et al.*, 1989). Rocks serve as reservoirs of phosphates, releasing them into water bodies. Additionally, bed sediments also contribute to phosphate levels in surface waters. Biological wastewater treatment processes often result in phosphate discharge into water bodies, while food waste from households, body wastes, animal excretory materials, and animal manures also add to phosphate pollution (Correll, 1998).

Emerging Sources and Trends of Phosphates

A new source of phosphates comes from household detergents used in washing clothes, which are drained into water systems (Neal *et al.*, 1999). Excessive use of phosphorus-rich fertilizers increases phosphate levels in agricultural runoff. Extensive growth of macrophytes can help remove a significant amount of phosphates from water bodies (Umerfaruq and Solanki, 2015). A study by Bhat *et al.* (2017) in Dal Lake across different seasons revealed trends similar to nitrate loading. Higher phosphate levels in summer and spring were attributed to increased activities during these seasons, including higher erosional activities in the catchment area and weathering of parent bedrocks in the lake's tributaries. Dry conditions, reduced water volume, and decreased agricultural activities explain lower phosphate levels in autumn. Based on these observations, Dal Lake's trophic status was predicted to be mesotrophic, though it could easily shift to eutrophic conditions under current circumstances. Conversely, Singh et al. (2008) reported that the hypolimnion region of Dal Lake is hypertrophic, similar to Tsokar Lake in Ladakh, while Mansar, Surinsar, and Tsomoriri Lakes in Ladakh are eutrophic.

Trophic Status and Pollution Impact on Anchar Lake

Anchar Lake in the Srinagar district of Kashmir has been assessed by Tahseen *et al.* (2018), revealing that the lake suffers from high nutrient and organic biomass loading, which boosts productivity but deteriorates water quality. This decline is exacerbated by increased urbanization and agricultural activities around the lake and its catchment area. Excessive fertilization of floating gardens contributes to algal blooms, with toxic red blooms reported by Khan (2000) harming native fish species. According to Bashir et al. (2017), chemical discharges from laboratories at the Sher-e-Kashmir Institute of Medical Sciences, despite treatment, negatively impact the lake and occasionally become toxic to aquatic organisms, compounded by municipal waste. The combination of toxic pollutants and climate change pressures poses a significant threat to other indigenous fish species, such as *Triplophysa kashmirensis,* mahseer fish (*Tor putitora*) from the River Tawi in Jammu, and *T. landacensis* (Ladakh loach) (Arjamand *et al.*, 2013).

3. Harmful Algal Blooms (HABs)

An algal bloom refers to a rapid increase in algal species in water, leading to discolouration due to their pigments (Foster, 2013). While algal blooms play a crucial role in aquatic ecosystems through their photosynthetic activity, influencing productivity in both natural and cultured environments, their unchecked proliferation can have significant ecological, economic, and health consequences. They can adversely affect fisheries, aquaculture, and tourism, and may impart an undesirable earthy or moldy taste (due to geosmin) to fish. Algal blooms often signal an imbalance in the ecosystem and are driven by various factors, including anthropogenic activities, climate change, and broader human impacts. Changes in watersheds, climate-induced temperature shifts, and other biological changes can also contribute to the development of these blooms (Watson *et al.*, 2015).

Eutrophication and Algal Blooms in the Dal Lake

Dal Lake, like many lakes in Kashmir, is predominantly eutrophic, as supported by existing literature. The lake faces challenges from harmful red algal blooms, alongside issues such as weed infestation, pollution, and various anthropogenic disturbances. These algal blooms diminish the lake's aesthetic appeal, impact plant and animal life, and are particularly prevalent in areas like Hazratbal and the Boulevard. Major contributors to these blooms include sewer discharge, plastic waste, stagnation, and a lack of local awareness. Additionally, the presence of houseboats, residential hamlets, and heavily fertilized floating gardens exacerbates the problem. Changes in the valley's temperature regime over the years have further negatively affected species diversity and spatial distribution.

Triggers of Eutrophication and Algal Blooms

The factors driving eutrophication and algal blooms are largely similar, with excess phosphates and nitrates playing a key role in both processes. High levels of carbon and residual sodium carbonate, which supply dissolved CO_2, enhance photosynthetic activity and promote algal growth. Due to their short lifespan, algae eventually die and decompose, creating a substantial organic load that consumes large amounts of oxygen. This process leads to hypoxic conditions, endangering the survival of aquatic organisms. Dal Lake, with an average depth of just 2.5 to 3 meters, is highly susceptible to algal blooms, particularly given the levels of pollution and nutrient enrichment caused by human activities. The lake experienced its first reported algal bloom in 1991, attributed to the Euglenoid species *Euglena rubra,* as observed by LAWDA. This bloom was largely due to eutrophication and sudden changes in water quality, exacerbated by high chloride and nitrate levels resulting from stagnant water, accumulation of free carbon dioxide, and nutrient enrichment.

In April 1998, a bloom of Cladophora caused the Nigeen basin's waters to emit a foul odor similar to rotten eggs, a result of untreated sewage. Kundanger (1999) also reported blooms of Volvocales and *Microcystis aeruginosa,* which turned the waters in Hazratbal and Nehru Park a lush green. Additionally, Lone *et al.* (2017) noted *M. aeruginosa* blooms in the Brein Laam area of Dal Lake. LAWDA's 2012-13 assessment found that cyanophyceae contributed to 50% and 80% of the total phytoplankton population at two different sites. The following year, these figures were 50.4% and 58%. The summer of 2015 saw the highest bloom populations, while the autumn of 2014 experienced the lowest, following the severe floods of September 2014.

Concerns with Invasive Bloom-Forming Algae

Invasive species of bloom-forming algae pose significant concerns as they exploit favourable conditions to invade ecosystems, leading to imbalances, genetic pollution, habitat destruction, and disruptions to aquaculture activities. This can severely impact the ecological balance. Understanding these invasive species and implementing management strategies are crucial to prevent their growth and spread. Algal toxins can cause serious health issues, including contamination of drinking water, which can lead to health problems in humans, massive fish kills, and animal deaths. These toxins and odorous metabolites might even evade standard water treatment processes, resulting in foul-smelling and flavoured drinking water along with health risks. Some toxins have been linked to illnesses such as paralysis and are suspected to contribute to liver cancer. Hepatotoxicosis can be caused by cyanobacterial peptides like microcystins and nodularins. Additionally, genera such as Oscillatoria and Anabaena can produce alkaloids that act as neuromuscular blocking agents, potentially causing fatal outcomes. Cyanobacteria also produce lipopolysaccharides/endotoxins, which can lead to allergic reactions. Invasive bloom-forming algae are a major concern because they can capitalize on favourable conditions to invade ecosystems, causing imbalances, genetic pollution, habitat destruction, and interference with aquaculture activities. This disrupts ecological stability. To manage these risks effectively, it's crucial to understand these invasive species and implement strategies to control their growth and spread.

Algal Bloom Control Measures

Controlling algal blooms involves several strategies. Real-time monitoring of water quality is essential, utilizing both direct and indirect methods such as measuring chlorophyll-a, phycocyanin, temperature, and dissolved oxygen. This monitoring helps predict potential algal blooms. Copper algicides are not an ideal solution due to the risk of copper toxicity and the potential for cyanobacterial lysis, which can release harmful microcystins. Biological control, using predatory bacteria from the genus *Bacteroides*, offers a reliable alternative. Ultrasonic algae control is an eco-friendly method that eliminates the need for chemicals. For long-term solutions, upgrading sewage treatment plants with advanced technologies is crucial. Wastewater from houseboats and nearby settlements should be directed to these treatment plants to prevent the discharge of untreated waste. Reducing anthropogenic impacts in the catchment area, such as waste disposal, excessive fertilizer use, grazing, and deforestation, is also necessary. Increasing public awareness about the consequences of these activities is important. Promoting organic farming can serve as an alternative to reducing nutrient runoff. An example of the negative impact of human activity is the nutrient enrichment of the lake in the Kralsangruip to Laam region due to the disposal of cow dung. This highlights the need for greater awareness among local communities. Collaborative research involving experts from LAWDA, SKUAST-K, and the University of Kashmir is essential for the effective management and mitigation of algal blooms.

4. Impact of Tourism

Tourism is a key economic driver for the UT, with water bodies playing a crucial role in attracting visitors from around the globe. Major attractions include Dal Lake, Wular Lake, and various streams, lakes, waterfalls, and rivers. Activities such as angling and fish watching are emerging as significant sources of income and employment. Revenue from fishing licenses supports fish and wildlife management at all government levels. Recreational fishing expenditures also contribute to local and regional economies, especially where fisheries are maintained in pristine or near-pristine conditions. Angling, viewed as a form of ecotourism, has minimal environmental impact while aiding in system preservation and sector support. In the fiscal year 2015-16, the Department of Fisheries (DoF) issued 1,118 trout permits, with 1,068 granted to non-citizens and 50 to citizens. Due to the substantial revenue generated from tourism, the department often prioritizes tourism development over the needs of local fishers who rely on local species for their livelihoods (Qureshi *et al.*, 2016).

Trends and Challenges in Fishing Licenses and Sport Fisheries

There is an increasing trend of acquiring fishing licenses primarily to receive compensation during the fishing ban season from March to May when snow trout are breeding. This practice means the actual number of active fishers is significantly lower than reported. According to J&K Bank, the sector has faced major setbacks due to poaching in trout streams. In response, new trout streams have been established in districts such as Kupwara, Pulwama, Baramulla, Anantnag, and Budgam. Recreational parks in Gulmarg and Sonamarg aim to attract foreign tourists. The Mahseer (*Tor* spp.) is a key species in sport fishing, but it has faced significant challenges due to obstacles in migration caused by dam and barrage constructions across rivers. Notably, the Golden Mahseer (*Tor putitora*) has disappeared from the River Jhelum, affected by the Mangla Dam further downstream in Pakistan.

Impact of Houseboats and Sanitation Issues

Houseboats are a significant tourist attraction on Dal Lake. The J&K Tourism Department has limited the number of houseboats on Dal and Nigeen lakes to 910 (TOI, 2021). In addition to houseboats, the lakes also feature doonga and shikara boats. While houseboats are equipped with water and electricity connections, there is a lack of proper moorage systems, and waste is often disposed of untreated into the lake. This contributes to serious problems such as eutrophication, algal blooms, and threats to aquatic life. Similarly, lakeshore communities around Lake Wular face significant sanitation challenges. Many residents lack proper sanitation facilities, leading to the disposal of wastewater directly into the lake, which poses health risks. Currently, fewer than 2% of households have access to safe sanitation facilities, and 42% rely on untreated water from Lake Wular for their domestic needs (Nazir, 2018).

Decline of Euryale ferox *(Prickly Waterlily)*

Euryale ferox, commonly known as makhna, is declining due to human activities. Its prickly nature and its interference with the growth of *Nelumbo nucifera* (Nadru), a widely used food source, have led to its eradication. The cultivation of Nelumbium, Trapa species, and Azolla, which form dense layers, can also impede its growth. Pollution further exacerbates its decline. In Dal Lake, water skiing in the Nigeen and Nehru Park basins, where this weed was considered a nuisance, led to its large-scale eradication. Similarly, boat pullers in Dal Lake removed the weed to prevent disruption during shikara rides. This demonstrates how prioritizing tourism and income-generating activities led to the compromise of this aquatic resource. Recognizing the threat to this plant, efforts were made to germinate its seeds in laboratories and transplant them to favourable sites in Manasbal, Anchar, and Dal Lakes. These efforts have successfully increased the plant's population and raised awareness among local communities about its preservation. Protecting this resource is crucial as it has medicinal value, can be consumed directly, and used to make flour for bread (Kak, 2010).

Impact of Floating Gardens

Floating gardens and the local lake-dwelling population, known as Mir Behris, are distinctive features of Dal Lake. These inhabitants contribute to government revenue through their activities, including cultivating vegetables in floating gardens and consuming lake resources. However, their growing numbers have placed increasing stress on the lake's ecosystem. Pollution from their settlements and the use of fertilizers on the floating gardens have exacerbated this issue. The presence of floating gardens has also obstructed waterways and reduced the area of clear water. To create stable land for their gardens, lake dwellers often dump mud and weeds into the lake, transforming parts of it into solid masses. This practice is particularly prevalent along the western shore behind Dole Demb. Lake dwellers reside in three main types of accommodations: peripheral areas of the lake, Doonga boats, and hamlets. Approximately 7,500 people are living in houseboats and around 100,000 in hamlets. Additionally, the influx of tourists further pressures the lake's ecosystem.

5. Dams, Barrages and Weirs

The construction of the Mangla Dam in Pakistan, located downstream on the River Jhelum, exemplifies how anthropogenic interference can endanger species. This dam, which lacks a fish ladder, has blocked the migratory route of the mahseer fish, preventing them from reaching their breeding grounds in Kashmir. Although the Department of Fisheries (DoF) has recognized the severity of the issue and established a hatchery in Anjj, the damage has already been done. Such river valley projects not only disrupt migratory patterns but also cause significant morphological changes to river systems. These changes include alterations in river shape and bed profile, submersion of gravel zones or riffle sections, destruction of riparian vegetation, and shifts in trophic regimes (Arjamand *et al.*, 2013). The construction of embankments in the Jhelum basin, particularly between Asham and Baniyar, has prevented the dispersion of silt, leading to its accumulation in Wular Lake. These embankments have hydrologically isolated and fragmented the marshes from the lake. Satellite imagery shows a noticeable northward shift in the river course after Baniyar, attributed to heavy silt deposition in the inlet. The Ningli Barrage Project for Wular Lake has significantly reduced the lake's water-holding capacity. Over the past 30 years, the lake's capacity has diminished by about 91.56 million cubic meters (MCM) at the maximum water level of 1579 meters above sea level (amsl). This loss represents approximately one-fifth of the lake's capacity and corresponds to an annual sedimentation rate of about 2470 acre-feet.

Effects of Siltation: The Case of Wular Lake

Wular Lake, the largest freshwater lake in India, located in the Bandipora district of Kashmir, is significantly impacted by siltation. Over time, siltation has caused the lake to shrink, affecting its flora, fauna, and the local communities that rely on it for their livelihoods. The lake plays a crucial role in regulating the flow of the River Jhelum by storing excess water during peak summer months and releasing it during the drier periods. Designated as a Ramsar site, Wular Lake's importance is further underscored by its role as a key habitat for migratory birds along the Central Asian Flyway, contributing to its rich biodiversity. Additionally, a substantial population dependent on the lake's fishery lives along its edges, benefiting from resources such as water chestnut, in addition to other traditional uses. Siltation in Wular Lake results from both natural and human-induced factors. The lake, situated in a tectonically active region, experiences natural erosion and seismic activity. However, anthropogenic activities significantly exacerbate the problem. The catchment area, characterized by natural erosion, has seen accelerated degradation due to increased human settlement. The growing population has encroached upon and deforested hill slopes, converting these areas into agricultural and horticultural land. These lands are ploughed across the contours, leading to severe erosion. Sediment is then transported to the lake basin via streams, channels, and gullies, with fertilizers from these lands further contributing to pollution. Private enterprises encroaching on the lake area have also contributed to its shrinkage and erosion. High grazing pressure from nomadic herders on high-altitude pastures exacerbates the erosion problem. According to CMAP, approximately 2000 hectares are classified as highly eroded, 2500 hectares as moderately eroded, and 4100 hectares as slightly eroded.

These human activities have disrupted the lake's equilibrium, posing threats to both the environment and the local communities. Deforestation in the lake's catchment area has led to a shortage of firewood for winter use. For instance, in Kuhnis village, located on the banks of Wular Lake, the forest line has receded by 0.8 kilometers over the past 30 years. In response, efforts have been made to address the issue by planting willow trees over 27.30 square kilometers in the lake's marshes and Rubinia on drier sites. However, these measures have faced resistance from locals concerned about potential damage to the marshes, which are vital resources for them. The lake's declining capacity to support fisheries and other resources has led to increased exploitation practices, such as using nylon nets with smaller mesh sizes. This has heightened the risk of poverty for local communities due to unreliable income sources. Additionally, transportation between villages via waterways has been hindered by siltation and the reduction in water spread area. The lake's biodiversity is also suffering, with native fish species and waterbird populations dwindling, further contributing to the economic hardships faced by residents. Additionally, stone quarries operating year-round in the lake's direct catchment release loose detritus during the monsoon season. Mining activities for sand and bajri in the River Madhumathi basin further contribute to siltation. Consequently, the lake's water body has decreased from about 91% to 75%, and marshy areas have reduced from 66% to 11%, reflecting an overall change of more than 40% due to increased plantation, agriculture, and settlement activities.

The region is increasingly vulnerable to flooding as well. The negative impacts of human activities on Wular Lake are evident and require urgent and serious intervention. The state government established the Manasbal and Wular Development Authority (MWDA) under the Department of Tourism, which identified 21 action points. However, the authority has struggled with coordination, long-term objectives, and a clear vision. Similarly, a wetland management plan was developed in 2007 and implemented in 2011 by the Wular Conservation and Management Authority. Nevertheless, there is a need for more effective and reliable initiatives from the relevant authorities and departments. Lake Manasbal is another casualty of siltation, primarily caused by stone quarrying and kiln operations near its Kondabal side. Despite a ban imposed by the Department of Tourism on these activities due to concerns about air quality, they continue to occur. Siltation is the main consequence, leading to reduced light penetration, decreased primary productivity, and increased BOD (Urich, 2002). The kilns contribute high levels of limestone-rich sediments (Rashid *et al.*, 2013). Additionally, the quarrying and transportation of gypsum generate high noise levels, which disturb both the lake's biodiversity and local communities (Magray and Jan, 2014). Quarrying also removes scrub and other vegetation in hilly areas, increasing vulnerability to small tremors and soil destabilization (White and White, 1995; Darwish *et al.*, 2010). Similarly, Dal Lake faces siltation threats from developments on the foothills of Shankar Acharya and Zabarwan, and encroachments like the para-military garrison in its catchment area (Jeelani *et al.*, 2008).

6. Wetlands Destruction

The scenic wetlands of Kashmir are not only major tourist attractions but also crucial for their ecosystem services. They support agriculture, fisheries, and related activities through their high productivity and provide essential hydrological functions such as groundwater recharge, flood mitigation, stream flow regulation, and carbon sequestration (Romshoo and Rashid, 2014). However, anthropogenic activities have severely threatened these valuable systems. Many wetlands are now considered wastelands due to the accumulation of various waste materials (Bano *et al.*, 2018). In Kashmir Valley, the trend of converting agricultural lands into urban areas further exacerbates the problem (Rashid and Romshoo, 2013; Rashid *et al.*, 2017). Consequently, pollution, nutrient overload, and reduction in water spread areas have significantly degraded these aquatic systems, including wetlands.

The wetlands have lost their ability to buffer floodwaters due to encroachment and infrastructure development, leaving the region more susceptible to flooding (Romshoo *et al.*, 2017). The Brari Nambal marshy lagoon, which previously channeled excess water to Khushalsar Lake and served as a waterway connecting Dal Lake to Anchar Lake, has been converted into a motorable road, disrupting its hydrology (Wani *et al.*, 2014). Additionally, partially treated sewage from the STP on the southern side has further compromised the water quality and aquatic life in this wetland. The twin lakes of Gilsar and Khushalsar, located to the northwest of Srinagar, have been subjected to encroachment, landfilling, illegal construction, and sewage disposal, resulting in severe weed infestation (Nissa and Bhat, 2016; Chowdhury, 2017). In Hokersar Wetland, a Ramsar site, the aquatic macrophytes *Acorus calamus, Euryale ferox,* and *Nelumbo nucifera* have disappeared due to siltation and reduced water depth, leading to an infestation of *Azolla* spp., *Salvinia natans,* and *Menyanthes* spp. (Bano *et al.*, 2018). The Lakes and Waterways Development Authority (LAWDA) in Srinagar is responsible for the conservation and management of wetlands.

However, their failure to adequately address these responsibilities is evident from the deteriorating condition of the wetlands. A comprehensive wetland management plan should include several steps to protect, restore, and manage these ecosystems, ensuring they continue to provide valuable functions and support sustainable use (Walters, 1986).

7. Fisheries Exploitation

The overexploitation of fishery resources significantly contributes to the decline of fisheries in Kashmir. Anthropogenic activities have led to a collapse in the diversity, capacity, and productivity of water bodies, reducing their ability to support local fishers' income needs. As a result, individuals turn to increasingly exploitative fishing practices, including the use of smaller mesh sizes and illegal methods such as explosives, in an attempt to secure a sufficient harvest. However, these practices exacerbate the problem rather than provide a sustainable solution. Overfishing has been reported in several Kashmir water bodies, including Dal and Wular lakes. According to Sunder and Subla (1984) and Yousuf et al. (2006), fishing effort serves as an indicator of the fish population structure within a water body. Dal Lake, in particular, has experienced a decline in fish populations due to overexploitation and overfishing, compounded by urbanization, pollution, and habitat and breeding ground destruction (Shafi and Yousuf, 2012).

8. Effects of Climate Change and Global Warming

Kashmir's rivers and water bodies rely on Himalayan snowmelt and glaciers for their perennial flow. However, rising temperatures due to global warming and greenhouse gas emissions are causing these glaciers to retreat, putting the rivers at risk of reduced water supply (Jeelani *et al.*, 2008). The Climate Profile document of Jammu and Kashmir reports a temperature increase in the region, with Kashmir experiencing a rise of 1.45°C and Jammu 2.34°C over the past two decades. Temperatures in Kashmir Valley have risen by 0.5°C per year, while Jammu has seen an increase of 0.8°C per year. Past interglacial periods were at least 2°C warmer than current conditions, aligning with predictions of anthropogenic warming by 2100 (Folland *et al.*, 2001). As cold-blooded organisms, fish are directly affected by water temperature. Higher temperatures lead to increased metabolism and feeding rates, and fish reach maturity earlier but tend to be smaller in size. Low dissolved oxygen (DO) levels, decomposing macrophytes, and vegetation contribute to adverse conditions for trash fish like Gambusia. However, experts at SKUAST-K attribute fish kills primarily to pollution. Rising temperatures are accompanied by reduced precipitation and snowfall (Dimri and Kumar, 2008; Dar *et al.*, 2014). Reduced snowfall leads to early snowmelt, resulting in an earlier spring and subsequently higher surface temperatures (Dunne *et al.*, 2003; Zaz and Romshoo, 2013). Furthermore, decreased precipitation can lead to drought conditions.

The National Institute of Hydrology (Roorkee) reports that glaciers in Ladakh, Zanskar, and the Great Himalayan ranges of Jammu and Kashmir are generally receding, with changes in glacier volume ranging from 3.6% to 97%, and most glaciers showing a degradation of 17%–25%. The 23 km stretch of the Drang-Drung glacier in Zanskar Valley is impacted by western disturbances, and the Siachen glacier has also experienced a reduction in glacial area. These conditions indicate that Kashmir's water resources may face potential scarcity, and its flora and fauna could struggle due to rising temperatures and environmental degradation. Cold-water fishes, being highly sensitive to changes in water temperature and environmental conditions, serve as crucial ecological indicators of climate change in temperate regions. Many endangered Himalayan cold-water fish species are particularly vulnerable to climate change. Ecological models predict significant losses of temperate fish species due to climate change's potential to reduce fish habitat and biodiversity, as these species are already at the upper limits of their thermal range in many areas (Mohseni *et al.*, 2003).

9. Threats to Migratory Birds and Water Birds

Kashmir, home to key migratory bird habitats such as Hokersar Wetland and Wular Lake, provides essential breeding sites for water birds. However, there is a notable lack of comprehensive mapping of these breeding sites, prey availability, habitat relationships, and preferences. Human activities pose significant threats to waterbirds, primarily through hunting and poaching. Decreased wetland areas, inadequate food sources, and reduced plant cover have led to declining water bird populations. The absence of formal conservation status for many sites exacerbates the problem, as large numbers of geese and ducks are hunted in unprotected areas. This has forced the birds to relocate to protected areas like Haigam during the day and return at night. Additionally, the removal of eggs and chicks from nesting water birds leads to failed breeding attempts. Urbanization, pollution, and the loss of feeding and breeding grounds due to heavy grazing, alongside the transformation of open water areas into vegetation-dominated spaces, further threaten these birds. Overfishing within their habitats reduces the availability of fish and invertebrates, impacting both resident and migratory birds. These combined factors disrupt the delicate balance of aquatic ecosystems and lead to a decline in biodiversity due to human interference.

Conclusion

Cold-water habitats are inherently low-productivity systems and are highly sensitive to human impacts. Any negative influence on these environments can have severe repercussions. Fisheries, which are crucial for the income and nutrition of local communities, will be adversely affected. Pollution and eutrophication in lakes threaten both the aquatic flora and fauna and pose health risks to individuals who rely on these water sources for various activities. These issues underscore the urgent need for measures to mitigate detrimental activities. While it is essential to enhance fish production, boost state revenue through tourism, and improve the living standards of local populations, these goals must not be pursued at the expense of jeopardizing natural resources and causing environmental degradation. Achieving these goals requires a sustainable approach that involves active participation from primary stakeholders, particularly the fishers who rely on these resources. Educating the local population about the significance of protecting water resources and the methods to do so is crucial. Sewage treatment plants must operate efficiently and reliably to prevent nutrient loading and pollution from harming water bodies. Promoting organic farming in the catchments of lakes, rivers, and water bodies will offer long-term benefits. Additionally, it is essential to regulate the overexploitation of resources such as fish, water chestnut, Nelumbo, and prickly water lily. Sand mining, quarrying, deforestation, and overgrazing must also be controlled to maintain healthy and hospitable aquatic ecosystems. Raising awareness about the crucial role of migratory and water birds in sustaining biodiversity and maintaining ecosystems is essential. Educating people on their importance may encourage voluntary cessation of hunting and poaching. Designating vulnerable areas as protected waters would be a positive step forward. Management and conservation authorities must enhance coordination to effectively address the threats facing water bodies, flora, and fauna. Implementing stringent regulations to safeguard these resources and revitalizing deteriorating areas is crucial. Legislation such as the Jammu and Kashmir Water Resources (Regulation & Management) Act, 2010, exemplifies such necessary measures (Dar and Buch, 2022).

References

Arjamand, S., Dar, S. A., Desai, A. Y., Sayani, A. N., Yusufzai, S. I., Ashfaq, M., Bhola, D. V. and Fofandi, M. D. 2013. Reproductive biology of an endangered coldwater fish Golden Mahseer, Tor Putitora (Ham.) from Anji Mahseer Hatchery Reasi (J&K). IOSR Journal of Pharmacy, 3(10): 13-16.

Bano, H., Lone, F. A., Bhat, J. I. A., Rather, R. A., Malik, S., and Bhat, M. A. (2018). Hokersar Wet Land of Kashmir: its utility and factors responsible for its degradation. Plant Arch., 18:1905–1910.

Bashir, M., Chauhan, R., Mir, M. F., Ashraf, M., Amin, N., Bashir, S. A. and Tabasum, S. 2017. Effect of pollution on the fish diversity in Anchar Lake, Kashmir. International Journal of Fisheries and Aquatic Studies, 5(1): 105-107.

Bhat, R. A., Shafiq-ur-Rehman., Mehmood, M. A., Dervash, M. A., Mushtaq, N., Bhat, J. I. A. and Dar, G. I. 2017. Current status of nutrient load in Dal Lake of Kashmir Himalaya. Journal of Pharmacognosy and Phytochemistry, 6(6): 165-169.

Carlson, R. E. 1977. A trophic state index for lakes. Limnology and Oceanography. 22: 361-369.

Chowdhury, C. L. 2017. Greater Srinagar and Dal Lake integrated environmental project proposal–a review [In: Geomechanics and Water Engineering in Environmental Management, (Ed). R. N. Chowdhury (London, UK: Routledge)] pp: 233–266.

Correll, L. 1998. The role of phosphorus and eutrophication of receiving waters: A review. J. Environ. Quality. 27:261-266.

Dar, M. A. and Buch, M. A. 2022. Regulating water bodies in the Kashmir valley. India Water Portal.https://www.indiawaterportal.org/articles/regulating-water-bodies-kashmir-valley [Accessed on October 30, 2022].

Dar, R. A., Rashid, I., Romshoo, S. A., and Marazi, A. 2014. Sustainability of winter tourism in a changing climate over Kashmir Himalaya. Journal Environment Monitoring and Assessment, 186(4):2449-2562.

Darwish, T., Khater, C., Jomaa, I., Stehouwer, R., Shaban, A. and Hamze, M. 2011. Environmental Impact of Quarries on Natural Resources in Lebanon. Land Degradation and Development. 22(3), pp 345–358.

Department of Fisheries, Government of Jammu and Kashmir. 2022. History of the department. http://jkfisheries.in/History.htm [Accessed on October 30, 2022].

Department of Fisheries. Handbook on fisheries statistics. 2020. pp: 8.

Dimri, A. P., and Kumar, A. 2008. Climatic variability of weather parameters over the Western Himalaya: a case study. Proceedings of the National Snow Science Workshop, India. Snow and Avalanche Study Establishment.

Dunne, J. A., Harte, J. and Taylor, K. J. 2003. Subalpine meadow flowering phenology responses to climate change: Integrating experimental and gradient methods.

Fakayode, S. O. 2005. Impact of industrial effluents on water quality of the receiving Alaro River in Ibadan, Nigeria. Ajeam-Ragee. 10:1-13.

Folland, C.K., T.R. Karl, J.R. Christy, R.A. Clarke, G.V. Gruza, J. Jouzel, M.E. Mann, J. Oerlemans, M.J. Salinger and S.-W. Wang (2001). Observed Climate Variability and Change. In: Climate Change 2001: The Scientific Basis. Contribution of Working Group I to the Third Assessment Report of the Intergovernmental Panel on Climate Change [Houghton, J.T., Y. Ding, D.J. Griggs, M. Noguer, P.J. van der Linden, X. Dai, K. Maskell, and C.A. Johnson (eds.)]. Cambridge University Press, Cambridge, United Kingdom and New York, NY, USA, 881pp.

Foster, J. M. 2013. Lake Erie is dying again, and warmer waters and wetter weather are to blame. Climate Change Progress.

Fried, J. J. 1991. Nitrates and their control in the EEC aquatic environment. Ecol. Sc., 20:3-11.

Gopal, B. 2012. Coldwater system, human impacts and climate change. [In: DCFR Silver Jubilee Compendium on Coldwater Fisheries. (Eds). Sarma, D., Pandey, A., Chandra, S. and Gupta, S. K.] pp: 97-102.

IUCN. 2010. The IUCN Red List of Threatened Species. https://www.iucnredlist.org/species/166525/174786567 [Accessed on October 30, 2022]

Jeelani, M., Kaur, H. and Kumar, R. 2008. Impact of climate warming on the biodiversity of freshwater ecosystem of Kashmir, India. [In: Proceedings of Taal 2007: The 12th World Lake Conference. (Eds). Sengupta, M. and Dalwani, R.] pp: 1103-1109.

Joshi, P. K., Humayun, R., and Roy, P. S. 2002. Landscape dynamics in Hokersar wetland-an application of geospatial approach. J. Indian Soc. Remote Sens., 30:1–5.

Kak, A. M. 2010. Euryale ferox (Juwar) in Kashmir Lakes (J&K State), INDIA. Journal of Economic and Taxonomic Botany, 34:1-11.

Kausar, N., Mir, R. A., Jan, U., Shah, G. M and Mir, T. A. 2017. Comparative studies on breeding biology of Schizothorax esocinus and Schizothorax curvifrons in natural conditions. International Journal of Fauna and Biological Studies, 4(4): 04-07.

Khan, M. A. 2000. Anthropogenic eutrophication and red ride outbreak in lacustrine system of the Kashmir Himalaya. Acta Hydrochimica et Hydrobiologica, 28:95-100.

Khan, M. A., Shah, M. A., Mir, S. S., and Bashir, S. 2004. The environmental status of a Kashmir Himalayan wetland game reserve: aquatic plant communities and eco-restoration measures. Lakes Reserv. Res. Manage., 9:125–132.

Kulkarni, G. J. 1997. Water supply and sanitary engineering. 10th Ed. Farooq Kitabs Ghar. Karachi. pp: 497.

Lone, F. A., Shah, M. A., Dar, S. A., Taban, N., DharS. K., Ahmad, S. and Mir, I. A. 2022. Algal Blooms in Dal Lake of Kashmir Himalayan Valley: Causes, Impacts and Management Measures. [In: Transforming Agriculture for Self-Reliant India] pp: 297-314.

Lone, J. A., Lone, F. A. and Suseela, M. R. 2017. A review on Phycological studies with special emphasis on Kashmir Himalayan Valley: Algal Biodiversity. Annals of Plant Sciences, 6(12): 1879-1885.

Magray, A. A. and Jan, T. 2014. Conservation status of Mansbal Lake: a case study. International Journal of Scientific & Engineering Research, 5(3):310-312.

Mlitan, A. M., Abofalga, A. and Swalem, A. 2015. Impact of treated wastewater on some physicochemical parameters soil and its fungal content. Int. J. Env. Sc. Develop., 6(5):369-374.

Mohseni, O., Stefan, H. G., Eaton, J. G. 2003. Global warming and potential changes in fish habitat in U.S. streams. Climate Change, 59: 289–409.

Nazir, A. 2018. Study of anthropogenic impacts of siltation in Wular Lake (Jammu and Kashmir). International Journal of Advance Research in Science and Engineering, 7(4):2073-2086.

Neal, C., Jarvie, H. P. and Ogunchi, T. 1999. Acid-available particulate trace metals associated with suspended sediment in the Humber River: A regional assessment. Hydrological Process, 13:1117-1136.

Nissa, M., and Bhat, S. U. 2016. An assessment of phytoplankton in Nigeen Lake of Kashmir Himalaya. Asian J. Biol. Sci., 9:27–40.

Qureshi, N. W., Krishnan, M. and Sundaramoorthy, C. 2016. 'Fish for all' versus 'fish of choice' – growth, instability and stakeholders' responses for enhancing fish production in major lakes of Kashmir. Current Science, 110(8): 1495-1504.

Qureshi, N. W., Krishnan, M., Sundaramoorthy, C., Vasisht, A. K., Baba, S. H., Kumar, N. R., Sharma, R. 2013. Truncated Growth and Compromised Sustainability: The Case of Lake Fisheries in Kashmir. Agricultural Economics Research Review. 26: pp 57-66.

Raina, H.S. and Peter, T. 1999. Fish and fisheries in the Indian Himalayas Lake and reservoirs, In: Fish and Fisheries at higher altitudes, FAO Technical Paper No. 9 385.

Rapin, F., Blanc, P. and Corvi, C. 1989. Eutrophication trends in the water quality of the Rhode River. J Mar. Bio. Assoc. 54:825-855.

Rashid, I., and Romshoo, S. A. 2013. Impact of anthropogenic activities on water quality of Lidder River in Kashmir Himalayas. Environ. Monit. Assess., 185:4705–4719.

Rashid, I., Bhat, M. A., and Romshoo, S. A. 2017. Assessing changes in the above ground biomass and carbon stocks of Lidder valley, Kashmir Himalaya, India. Geocarto Int., 32:717–734.

Rashid, I., Farooq, M., Muslim, M. and Romshoo, S. A. 2013. Assessing the impact of anthropogenic activities on Mansbal Lake in Kashmir Himalayas. International Journal of Environmental Sciences, 3(6): 2052- 2063.

Romshoo, S. A., Altaf, S., Rashid, I., and Dar, R. A. 2017. Climatic, geomorphic and anthropogenic drivers of the 2014 extreme flooding in the Jhelum basin of Kashmir, India. Geomatics Natl. Hazards Risk, 9:224–248.

Romshoo, S. A., and Rashid, I. 2014. Assessing the impacts of changing land cover and climate on Hokersar wetland in Indian Himalayas. Arabian J. Geosci., 7:143–160.

Sarma, D., Akhtar, M. S. and Singh, A. K. 2018. Coldwater fisheries research and development in India. [In: Aquaculture in India (Eds). Tripathi, S.D., Lakra, W. S. and Chadha, N. K.] New Delhi, Narendra Publishing House, pp: 93-133.

Sehgal, K. L. 1989. Present status of exotic cold-water fish species in India. Proc. Asian Fish. Soc. Spl. Pub. No. 1.

Sehgal, K.L. 1999. Coldwater fish and fisheries in the Indian Himalayas: rivers and streams. p. 41- 63. [In T. Petr (ed.) Fish and fisheries at higher altitudes: Asia.] FAO Fisheries Technical Paper 385. FAO, Rome. 304 p.

Shafi, S. and Yousuf, A. R. 2012. Length-weight relationship and condition factor of Schizothorax niger (Heckel, 1838) Misra from Dal Lake, Kashmir. International Journal of Scientific and Research Publications, 3:1-3.

Singh, H. R. and Lakra, W. S. 2008. Cold water aquaculture and fisheries. New Delhi. Narendra Publishing House. pp: 22, 28, 29.

Singh, O., Rai, S. P., Kumar, V., Sharma, M. K. and Choubey, V. K. 2008. Water quality and eutrophication status of some lakes of the Western Himalayan region (India). [In: Proceedings of Taal 2007: The 12th World Lake Conference. (Eds). Sengupta, M. and Dalwani, R.] pp: 286-291.

Sunder, S. and Subla, B. A. 1984. Fish and fisheries of river Jhelum Kashmir. Zoological Orientalis, 1:34-39.

Swain, S., Felix, S., Anthony, C. and Chhandaprajnadarsini, E. M. 2017. Exotic fish species introduced in India and its impacts. Scientific India, pp: 27-29.

Tahseen, Abubakr, A. and Balkhi, M. H. 2018. Trophic status assessment of Anchar Lake, Kashmir. Journal of Pharmacognosy and Phytochemistry, 7(4): 1154-1156.

TOI. 2021. J&K authorities cap the number of houseboats at 910 for Dal and Nigeen lakes. https://timesofindia.indiatimes.com/travel/travel-news/jk-authorities-cap-the-number-of-houseboats-at-910-for-dal-and-nigeen-lakes/articleshow/81763633.cms. [Accessed on October 30, 2022]

Umerfaruq, Q. M. and Solanki, H. A. 2015. Physico-chemical parameters of water in Bibi Lake, Ahmedabad, Gujarat, India. J. Pollut. Eff. Cont., 3(1).

Urich, P. B. 2002. Land use in karst terrain: Review of impacts of primary activities on temperate karst ecosystems. New Zealand Department of Conservation. pp: 60.

Walters, C. J. 1986. Adaptive management of renewable resources. New York, NY: Macmillan

Wani, M. A., Dutta, A., Wani, M. A., and Wani, U. J. 2014. Towards conservation of world-famous Dal Lake-a need of hour. Int. Res. J. Eng. Technol., 1:24–30

Watson, S. B., Whitton, B. A., Higgins, S., Paerl H. W., Brooks, B.W. and Wher, J. D. 2015. Harmful Algal Blooms. [In: Freshwater algae of North America] Chapter-20. Elsevier Inc.

Wetlands International - South Asia. 2007. Comprehensive Management Action Plan for Wular Lake, Kashmir. pp: 8-58.

White, W. B. and White, E. L., 1995. Thresholds for soil transport and the long-term stability of sinkholes. [In: Karst geohazards - Engineering and Environmental Problems in Karst Terrane (Eds). Beck, B. F., and Pearson, F. M. Proceedings of the Fifth Multidisciplinary Conference on Sinkholes and the Environmental Impacts of Karst, A.A. Balkema, pp.73-78.

Yousuf, A. R.; Bhat, F. A. and Mahdi, M. D. 2006. Limnological features of river Jhelum and its important tributaries in Kashmir Himalaya with a note on fish fauna. Journal of Himalayan Ecology and Sustainable Development, 1:37-50.

Zaz, S. N., and Romshoo, S. A. 2013. Recent variation in temperature trends in Kashmir Valley (India). J. Himalayan Ecol. Sustain. Dev., 8.

16

Forward-Looking Perspective on Sustainable Management, Conservation Strategies & Research Opportunities for the Future of Fisheries in Kashmir Valley

Rinkesh Nemichand Wanjari[1], Tasaduq Hussain Shah[1], Zaheer Abass*[1], Hudisa Banoo[1], Dechan Chuskit[1] & Manabjyoti Barman[2]

[1]*Faculty of Fisheries (FoFy), Sher-e-Kashmir University of Agricultural Sciences and Technology of Kashmir (SKUAST-K), Rangil, Ganderbal, Jammu and Kashmir*

[2]*ICAR-Central Institute of Fisheries Education, Panch Marg, off Yari Road, Andheri (West), Mumbai, Maharashtra*

Abstract

Aquatic ecosystems across India have endured significant stresses over the past century due to unplanned developmental activities, leading to severe environmental degradation. These ecosystems, vital for supporting biodiversity and providing essential services, have been profoundly impacted by local disturbances and widespread alteration of various human activities. Waterways have been extensively modified for transportation routes, diverted for agricultural and municipal water needs, dammed for hydroelectric power, utilized as industrial cooling systems, and straightened for logistical convenience. These anthropogenic uses and exploitations have resulted in profound consequences, including declines in fisheries, increased incidences of floods and droughts, and loss of biodiversity. Some of these losses may be irreversible, highlighting the urgent need for conservation, restoration, and effective management of aquatic ecosystems. This Chapter underscores the critical importance of implementing strategies to safeguard these vital resources, aiming to restore ecological balance and ensure sustainable use for future generations.

***Keywords**: Aquatic ecosystems, Biodiversity, Fisheries, Conservation, Sustainable, Ecological balance*

Introduction

The global population is on track to surpass 9 billion by 2050, leading to a staggering 70% increase in the worldwide demand for food, feed, and fibre. This population growth is closely linked to significant shifts in lifestyle and consumption patterns, primarily driven by the accelerating pace of urbanization. As a consequence, it is anticipated that there will be a decline in the consumption of grains and pulses, coupled with a significant increase in the consumption of vegetables, fruits, meat, dairy, and fish. Aquaculture, a time-tested industry, stands out as a crucial source of high-quality protein for humanity (Nash, 2010; Gui *et al.*, 2018). As our global population continues to expand, the challenges of resource scarcity and environmental degradation are becoming increasingly evident. Land resources alone are no longer sufficient to support the evolving needs of our society. Consequently, our oceans are emerging as a new frontier for human sustenance and development, with fish taking on a pivotal role as a source of high-quality protein and a fundamental dietary staple (Baiden and Bissiri, 2007). Over the past few decades, aquaculture has risen to become the fastest-growing sector within agriculture. Since 2013, its production has even outstripped that of wild fisheries (FAO, 2020). This formidable challenge presents particularly alarming situations in developing countries where poverty and hunger persist (World Bank, 2014). In such a context, fisheries and aquaculture are anticipated to play a vital role in ensuring nutritional security (Subasinghe, 2014). According to the FAO, food production will need to rise by 70% by 2050 to meet the rising demand brought on by population growth and dietary changes (FAO, 2009). By effectively using the limited water resources, the fisheries industry may serve to address this issue and contribute significantly to nutrition and food security. A significant source of nutrition, food, income, and livelihood for millions of people worldwide continues to be the fishing and aquaculture industries (FAO, 2016). Additionally, fish contribute to global food security by providing up to 17% of the protein consumed by the world's population (Bennet *et al.*, 2018), and fishing is one of the most widely practised livelihoods worldwide (FAO, 2014). Although fish are the most diversified group of vertebrates, they are the second most threatened species after amphibians (Collen *et al.*, 2014).

Freshwater ecosystems have been among the most endangered habitats worldwide in recent decades. Freshwater researchers have become more concerned about the devastation of riverine environments, changes in aquatic ecosystems, rapid decrease in water quality, and considerable decline in the variety and abundance of stream fishes. The presence, distribution, abundance, and diversity of fishes in the aquatic ecosystem are influenced and shaped by some environmental factors (Whitfield, 1999; Blaber, 2008). The health of these freshwater ecosystems is evident in the well-being of the fish assemblage structure they support, and they are regarded as an essential component of human life along with their resources. Concerning both space and time, these fish assemblages continually change (Tang *et al.*, 2015). The extinction rates of freshwater fauna are higher than for many other ecosystems and significantly surpass historical background rates, making freshwater habitats some of the most threatened on Earth (Richter *et al.*, 1997; Leidy and Moyle, 1997). People depend on freshwater, and many cultures are quickly running out of this valuable resource (Vörösmarty *et al.*, 2010). Although human growth poses the biggest danger to freshwater ecosystems, the benefits that healthy freshwater ecosystems provide are directly essential to our communities and economies (Postel and Carpenter, 1997). These services include the provision of food and clean water, control over pollutants, flow patterns, and climate, assistance with nutrient cycling, and cultural services including leisure, travel, and spiritual advantages (Vörösmarty *et al.*, 2005). Of the 7% of global biodiversity that supports human existence and functioning, freshwater ecosystems are the richest and most promising (Dudgeon *et al.*, 2006; Reid *et al.*, 2019). Several stressors, such as a changing climate, species invasion, changes in land use, point source and diffused pollution, water extraction, eutrophication, growing hydropower, and rising resource consumption, etc., have put streams, rivers, lakes, riparian areas, ponds, and wetlands in grave danger (Flitcroft *et al.*, 2019; Cantonati *et al.*, 2020).

The WWF study of the Living Planet Index states that freshwater organisms are rapidly going extinct and experiencing population declines. According to the Living Planet Index (2016), there was an 81% decline in freshwater vertebrate species in 2014, by 2018, that number had increased to 83% (Grooten and Almond, 2018). Moreover, data from the IUCN red list indicates that there is a risk of extinction for both freshwater invertebrate and vertebrate species (Ricciardi and Rasmussen, 1999; Collen *et al.*, 2014). The freshwater ecosystem accounts for approximately 84% of the threatened megafauna that is located outside of the current protected area (Carrizo *et al.*, 2017). While the decline in freshwater organisms is concerning, it is not sufficiently addressed. The essential benefits that freshwater fisheries provide to biodiversity and food security are underappreciated (Beard *et al.*, 2011). As a component of a cohesive community, inland fisheries rely on biodiversity to maintain their resilience in the face of anthropogenic disturbance and natural variation. The notion of ecosystem health encompasses both biotic and abiotic elements, as well as nutrient and chemical cycles and the services that ecosystems offer to humans (Rapport *et al.*, 1998). Therefore, an aquatic ecosystem that is robust to disturbance and preserves characteristics of ecosystem structure and function, such as habitat, species composition, genetic diversity, and production at levels comparable to those observed in the absence of contemporary human activities, can be considered healthy and productive. On the other hand, one or more of these characteristics are degraded in disturbed ecosystems. Productive fisheries are dependent on healthy ecosystems, and maintaining the health of ecosystems is necessary to safeguard fisheries. Novel forward-looking approaches are required in addition to predictive models and other decision-support tools to support decisions on complex issues like climate change, sustainable development, and environmental management and planning. These issues are characterized by high levels of uncertainty about the best course of action and the potential effects those options may have on various socio-ecological system components (Tonn *et al.*, 2000).

The UN Decade of Ocean Science highlights the potential for innovation in "finding solutions to climate change but also to enhance and preserve the biodiversity of the ocean, sustainably feed the population, end poverty and inequality, and stimulate a thriving and sustainable ocean economy." Foresight is a systematic forward-looking process that anticipates future threats and opportunities influencing complex systems in the long term and contributes to developing adaptive strategies to guide decisions about the future. A sustainable blue economy requires knowledge, research, development, and innovation as its main cornerstones. It draws attention to the systemic and complex nature of an issue by analyzing this through multiple and often competing worldviews and by drawing attention to the factors in the external environment—these relate to political, economic, social, technological, legal and ecological factors—that influence the direction and speed of change. It has a practical, application, in enabling decision-makers and scientists to consider the impacts of policy options and to prioritize decisions and research directions under different future scenarios. This sector is vital for nutritional security, especially in developing countries. However, freshwater ecosystems, crucial for aquaculture, are among the most threatened habitats, facing pollution, climate change, and habitat degradation. Ensuring sustainable aquaculture practices is essential to meet future food demands and protect aquatic biodiversity.

Sustainable Management and Conservation Strategies

Aquatic Conservation Strategy

Despite making up only 0.01% of the Earth's surface, rivers, lakes, and wetlands are home to a disproportionately substantial share of the planet's biodiversity. There are an estimated 44,000 scientifically identified species of freshwater biota, and freshwater fish alone make up around 25% of all extant vertebrate species. Freshwater biodiversity is typically more vulnerable than terrestrial biodiversity, according to lists of endangered species. Numerous concerns threaten the biodiversity of freshwater ecosystems. These include the indirect effects of terrestrial activities including forestry, agriculture, industry, housing construction, and mining, as well as the direct effects of dams, invasive species, overfishing, pollution, stream channelization, water withdrawals, and diversion (Prasad *et al.*, 2024). The protection of freshwater biodiversity requires the development and use of conservation measures. According to Ramachandra *et al.* (2002), the goal of the aquatic conservation strategy is to conserve and preserve the ecological health of watersheds and aquatic ecosystems by:

1. The distribution, richness, and complexity of landscape- and watershed-scale characteristics must be preserved to safeguard the aquatic systems that are home to certain species, populations, and communities.
2. Preserve the interconnectedness of space and time both within and among watersheds. Wetlands, headwater tributaries, upslope regions, and flood plains are examples of lateral, longitudinal, and drainage network linkages. These lineages must offer physically and chemically clear pathways to places that are essential to aquatic and riparian-dependent species' ability to meet their life cycle requirements.
3. Preserve the species composition and structural diversity of plant communities in wetlands and riparian zones to supply sufficient amounts and distributions of coarse woody debris to support physical complexity and stability, as well as to provide adequate nutrient filtering, appropriate rates of surface erosion, bank erosion, and channel migration.
4. To maintain evenly dispersed populations of native plant, invertebrate, and vertebrate riparian-dependent species, habitat must be preserved.

Restoration of Aquatic Ecosystems

The deterioration of freshwater ecosystems by a range of stressors has risen logarithmically as a result of several human activities undertaken to meet the demands of a growing population. Because of this, a lot of aquatic habitats require significant restoration or remediation. Fischer *et al.* (2021) defines restoration as the "return of an ecosystem to a close approximation of its condition before disturbance" or the re-establishment of pre-disturbance aquatic functions and related physical, chemical, and biological characteristics. It is not accomplished by the discrete manipulation of separate components; rather, it is a whole process. The goal is to mimic a self-regulating, natural system that is ecologically integrated with the environment in which it exists. Rebuilding previous physical conditions, adjusting the soil and water chemically, and manipulating the environment biologically, including reintroducing extinct native flora and animals, are frequently necessary steps in restoration (Zerbe, 2022). Although the emphasis of these principles is on scientific and technical matters, community viewpoints and values must be taken into account, much as in any environmental management endeavor. Building the support required to start the project rolling and guarantee the long-term protection of the restored region may be achieved by coordination with the local population and organizations that may be impacted by it. Partnership with all parties involved can also provide beneficial resources, such as funding, technical know-how, and volunteer assistance with implementation and oversight (Ramachandra, 2001).

1. The cornerstone for maintaining biodiversity is the presence of existing, comparatively intact ecosystems, which supply the biota and other natural resources required for the restoration of damaged systems.
2. Restoration planning should take into account any irreversible changes in the watershed that may affect the system being restored, and focus on restoring its remaining natural potential.
3. Objectives set the direction for action and offer the benchmarks for success. With the area's natural potential, the selected objectives should be feasible from an ecological and socioeconomic standpoint, taking into account the resources at hand and the level of community support for the project.
4. Focus on feasibility taking into account scientific, financial, social, and other considerations
5. Restore native species and avoid non-native species. Many invasive species outcompete natives because they are adept at colonizing disturbed areas and lack natural predators or controls.
6. Monitor and adapt where changes are necessary. Monitoring before and during the project is crucial for determining if goals are being met. If they are not, "mid-course" adjustments should be made. Post-project monitoring will help identify if additional actions or adjustments are needed and can provide valuable information for future restoration efforts. This process of monitoring and adjustment is known as adaptive management. Monitoring plans should be cost-effective and technologically feasible, always providing information relevant to achieving the project goals.

Pollution Mitigation

Pollution mitigation in aquatic environments is essential due to the diverse and critical roles these ecosystems play in supporting biodiversity and human well-being. Aquatic pollution can originate from various sources, including industrial processes, agricultural activities, urban development, and improper waste disposal. To effectively mitigate pollution, comprehensive strategies are necessary, starting with robust regulatory frameworks that enforce pollution control measures and set stringent standards for pollutant discharges. Implementing these regulations requires collaboration between government agencies, industries, and communities to ensure compliance and accountability. Moreover, technological advancements play a pivotal role in pollution mitigation efforts. Innovative wastewater treatment technologies, such as advanced filtration systems and biological treatments, help remove contaminants before they enter water bodies. Monitoring and early warning systems also enable quick responses to pollution incidents, minimizing their impact on aquatic ecosystems (Emenike *et al.*, 2021) Public education and outreach programs are crucial in raising awareness about the importance of clean water and encouraging sustainable practices among stakeholders. Furthermore, sustainable land use practices, such as promoting natural buffers along waterways and implementing green infrastructure in urban areas, can help reduce runoff and pollution loads entering aquatic systems. Restoration of degraded habitats and ecosystems also contributes to pollution mitigation by enhancing natural filtration processes and supporting biodiversity recovery (Häder *et al.*, 2020). Ultimately, successful pollution mitigation in aquatic environments requires a multi-faceted approach that integrates science-based policies, technological innovations, community engagement, and proactive conservation measures. By prioritizing pollution reduction and sustainable management practices, we can safeguard water quality, protect aquatic species, and ensure the resilience of aquatic ecosystems for future generations.

Invasive Species Management

The unregulated introduction of alien species poses a major threat to biodiversity. Existing mechanisms to regulate the introduction of invasive alien species are weak. Many alien species have entered the country without clearance from authorities, and more continue to arrive through porous borders (Abass *et al.*, 2024). Additionally, keepers and hobbyists often discard ornamental fish into open waters when they no longer want them or when the fish outgrow their aquariums. Effective enforcement, risk assessment, and monitoring mechanisms are essential to regulate the entry, culture, sale, and disposal of these species (Raghubanshi *et al.*, 2005). Recently, the capture of an exotic fish species, the Alligator Garfish, in Dal Lake has raised concerns among scientists. This species is non-native to the area, and its introduction could potentially disrupt the delicate balance of the lake's flora and fauna. Non-native species often lack natural predators in their new environment, allowing them to outcompete native species for resources such as food and habitat. This competition can lead to declines in native species populations, affecting the entire ecosystem. Furthermore, exotic species may introduce new diseases or parasites that native species are not adapted to, further threatening their survival. The presence of Alligator Garfish in Dal Lake highlights the importance of preventing the introduction of non-native species and the need for stringent monitoring and management practices to protect eco-fragile environments like Kashmir Valley.

1. There is a need to create a comprehensive database on exotic or invasive alien species and their invasion pathways. Strategies for managing or controlling these species must be developed based on risk assessment. Additionally, quarantine measures must be strengthened to prevent their spread.
2. State and Union Territory governments must rigorously enforce existing regulations by actively intervening to curb the illegal introduction, culture, and sale of species prohibited by regulatory agencies.
3. Voluntary guidance should be provided to prevent the unintentional introduction of Invasive Alien Species associated with the trade of live organisms, including pets, aquarium species, terrarium species, live bait, and live food.
4. Silver carp, common carp, and grass carp should be restricted or discouraged in favor of native species in aquaculture within the Himalayan States.
5. Promote public awareness about the risks of releasing non-native species into natural water bodies.

Draft to Law: The Process of Environmental Policy Ratification

To ensure that the lotic and lentic system is effectively conserved and managed, policymakers should obtain comprehensive inventory data about the physical, chemical, and biological aspects of the system. The data collected for a given river should be coherent and connected from its head to its mouth (i.e. its basin). To adequately make conclusions on the current state of conservation, functional role, and energy dynamics of allochthonous creatures, floodplain areas, and riparian vegetation (Srivastava, 2024). This kind of activity also aids in managing the river ecosystem's fragmented restoration techniques. It is recommended that policymakers focus on fundamental research projects conducted by research institutes in conjunction with "Infield" experimentation investigations that use practical application ways ranging from upstream to downstream. A river ecosystem's "In- situ" correct spectrum of pollution impacts, ecological circumstances, gaps in restoration plans, etc. is provided by this kind of beginning (Srivastava, 2024). Effective ratification and implementation of environmental policies are essential for protecting the environment and ensuring a sustainable future. Committees should prioritize long-term studies to assess the accuracy and ecological implications of restoration targets to manage policies effectively. It will make it possible for us to combine policy and science into a single decision-making process. Furthermore, as the environment and society are changing and rigid policies may become outdated, management needs to be adaptable and dynamic with the ability to deal with ongoing uncertainty (Green *et al.*, 2015). The capacity of numerous creatures to adapt, disseminate, and endure in the increasingly climatically favourable environment is being impacted daily by the changing climate. In the current environment, rules mandate the migration of specific individual species to a designated place for these creatures to successfully regulate the chance of survival and persistence (Dudgeon, 2006). Policymakers should work closely with researchers to choose species during migration, carefully considering any negative impact on native species. Furthermore, it is recommended that the policy guidelines undergo frequent updates to improve both the scientific foundation and the socio-economic outlook (Reid *et al.*, 2019).

Community-Based Management

We believe in the power of involving local communities in resource management. Our projects frequently incorporate community-based approaches, engaging local populations in decision-making processes and sustainable practices. This includes education and training programs on sustainable fishing, water conservation, and habitat protection. By empowering communities, we ensure the long-term sustainability of conservation efforts and foster a sense of stewardship towards aquatic ecosystems (Moore *et al.*, 2024).

Future Prospects and Way Forward

In navigating the future of aquatic environment management, several key strategies emerge as critical pathways forward. First and foremost is the integration of sustainable practices across all facets of aquatic resource utilization, from fishing and aquaculture to water quality management and habitat restoration. This integration hinges on robust regulatory frameworks that prioritize conservation and biodiversity preservation while promoting ecosystem resilience in the face of climate change impacts. Technological advancements, including the use of AI-driven monitoring systems and data analytics, will be pivotal in enhancing our understanding of aquatic ecosystems. These technologies enable real-time data collection, analysis, and predictive modelling, empowering decision-makers to implement adaptive management strategies effectively (Guo and Shahbaz 2024; Abatan *et al.*, 2024). By leveraging these tools, we can proactively address emerging threats such as invasive species introductions, habitat degradation, and pollution incidents. Furthermore, community engagement and stakeholder participation are essential for fostering local stewardship and ownership of conservation efforts. Empowering communities through education, capacity building, and inclusive decision-making processes ensures that management strategies are locally relevant and sustainable over the long term. Collaborative partnerships between governments, NGOs, academia, and industry will also be crucial in pooling resources, expertise, and innovative solutions to tackle complex challenges facing aquatic environments. Looking ahead, investment in research and development, coupled with strategic investments in infrastructure and institutional capacity building, will underpin effective management of aquatic resources (Rapport *et al.*, 1998). This includes expanding monitoring networks, enhancing enforcement of regulations, and promoting eco-certification schemes to incentivize sustainable practices. By adopting a holistic and forward-thinking approach, we can safeguard aquatic biodiversity, support livelihoods dependent on these resources, and ensure the resilience of aquatic ecosystems for future generations.

AI-Driven Species Identification and Population Assessment

Artificial Intelligence (AI) has transformed the study and management of aquatic biodiversity. Our AI-driven systems analyze extensive data from underwater cameras and sensors to identify and count various species. This technology enables real-time biodiversity monitoring, allowing for quick assessments of ecosystem health and the effectiveness of conservation efforts. Additionally, it plays a crucial role in detecting changes in species populations, serving as early indicators of environmental stress or degradation.

Sustainable Fishing Technologies

Sustainable fishing is essential for preserving healthy aquatic ecosystems and ensuring the long-term viability of fish populations. Our organization promotes the use of sustainable fishing technologies and practices, such as eco-friendly fishing gear that minimizes bycatch and habitat destruction. We advocate for selective fishing techniques and enforce fishing quotas based on scientific research. Additionally, we support the use of AI-assisted tools for real-time monitoring of fish populations, enabling fishermen to target abundant species and avoid overfishing vulnerable stocks.

Electronic Tagging and Tracking for Wildlife Monitoring

Understanding the movements and behaviors of wildlife is crucial for protecting aquatic biodiversity. We use electronic tagging and tracking technologies to monitor aquatic species, especially those that are threatened or endangered. By attaching small electronic devices to animals, we collect data on their location, migration patterns, and behavior. This information is invaluable for conservation efforts, helping us identify critical habitats, understand breeding patterns, and monitor the impacts of environmental changes on these species.

Sensor Networks for Real-time Monitoring

In aquatic ecosystem management, real-time data collection is essential. Our extensive sensor networks, deployed in various aquatic environments, continuously monitor key parameters such as water temperature, pH levels, dissolved oxygen, turbidity, and the presence of specific pollutants. These IoT-enabled sensors provide a continuous data stream, offering insights into ecosystem health and enabling rapid responses to adverse changes. This network is crucial for the early detection of pollution incidents and for monitoring long-term trends in aquatic ecosystem health.

Data Analytics for Ecosystem Health Analysis

We analyze vast amounts of data collected from sensor networks, satellites, and drones using advanced data analytics techniques. This involves employing machine learning algorithms and predictive models to understand complex ecosystem dynamics and identify patterns in environmental changes. These insights guide our conservation strategies, enabling us to proactively address issues and tailor our interventions more effectively.

Certification and eco-labeling

Certification and eco-labeling are crucial tools for advancing sustainable fishing and aquaculture practices. Establishing capacity and institutional mechanisms is urgently needed to support these efforts. Market promotion and incentives should prioritize fish produced or captured using biodiversity-friendly methods.

Conclusion

Aquatic ecosystem conservation and management require collaborated research involving natural, social, and inter-disciplinary study aimed at understanding the various components, such as monitoring of water quality, socio-economic dependency, biodiversity, and other activities, as an indispensable tool for formulating long-term conservation strategies. This requires multidisciplinary-trained professionals who can spread the understanding of the ecosystem's importance at local schools, colleges, and research institutions by initiating educational programs aimed at raising the levels of public awareness and comprehension of aquatic ecosystem restoration, goals, and methods. Actively participating schools and colleges in the vicinity of the water bodies may value the opportunity to provide hands-on environmental education which could entail setting up laboratory facilities at the site. Regular monitoring of water bodies (with permanent laboratory facilities) would provide vital inputs for conservation and management

References

Abass Z., Wanjari R.N., Barman M., Ronad S. and Hussain A. (2024). The Ecological Influence of Exotic Fish Species in India AgriGate- An International Multidisciplinary e-Magazine volume 4 issue 04 (ISBN: 978-81-965582-91):126:133pp

Abatan, A., Adeyinka, M. A., Sodiya, E. O., Jacks, B. S., Ugwuanyi, E. D., Daraojimba, O. H., and Lottu, O. A. (2024). The role of IT in sustainable environmental management: A global perspective review. International Journal of Science and Research Archive, 11(1), 1874-1886.

Baiden, G., and Bissiri, Y. (2007). High bandwidth optical networking for underwater untethered telerobotic operation. In OCEANS 2007 (pp. 1-9). IEEE.

Beard T.D., Arlinghaus R., Cooke S.J., McIntyre P.B., De Silva S., Bartley D., and Cowx I.G. 2011. Ecosystem approach to inland fisheries: research needs and implementation strategies. Biol. Lett. 7: 481–483.

Bennett, A., Patil, P., Kleisner, K., Rader, D., Virdin, J., Basurto, X., Fluet-Chouinard, E., Funge-Smith, S., and McIntyre, P. B. 2018. Contribution of fisheries to food and nutrition security. Proceedings of the National Academy of Sciences of the United States of America, 115(29): 46.

Blaber, S.J. 2008. Tropical estuarine fishes: ecology, exploitation and conservation. John Wiley & Sons.

Caudill, J. Chilton, E.M. Douglas, M. Meybeck, D. Prager, P. Balvanera, S. Barker, M. Maas, C. Nilsson, T. Oki, and A. Reidy. Island Press, Washington, DC. pp. 165–207.

Collen, B., Whitton, F., Dyer, E.E., Baillie, J.E., Cumberlidge, N., Darwall, W.R., Pollock, C., Richman, N.I., Soulsby, A.M. and Böhm, M. 2014. Global patterns of freshwater species diversity, threat and endemism. Global ecology and Biogeography, 23(1): pp.40-51.\

Emenike, E. C., Iwuozor, K. O., and Anidiobi, S. U. (2021). Heavy metal pollution in aquaculture: sources, impacts and mitigation techniques. Biological Trace Element Research, 1-17.

FAO, (2020). The State of World Fisheries and Aquaculture 2020. State World Fish. Aquac. 2020.

FAO, 2009. Food and Agriculture Organization of the United Nations. How to Feed the World in 2050—High Level Expert Forum, Rome 12–13 October, 2009.

FAO, 2014. The state of world fisheries and aquaculture. 243 pp. Retrieved 14 Feb 2024, from http://www.fao.org/3/a-i3720e.pdf.

FAO, 2016. Food and Agriculture Organization of the United Nations, State of world fisheries and Aquaculture 2016, Contributing to food security and nutrition for all, FAO Rome, pp. 200.

Fischer, J., Riechers, M., Loos, J., Martin-Lopez, B.,and Temperton, V. M. (2021). Making the UN decade on ecosystem restoration a social-ecological endeavour. Trends in ecology & evolution, 36(1), 20-28.

Gui, J.F., Tang, Q., Li, Z., Liu, J. and De Silva, S.S. eds., (2018). Aquaculture in China: success stories and modern trends. John Wiley & Sons.

Guo, X., and Shahbaz, M. (2024). The existence of environmental Kuznets curve: Critical look and future implications for environmental management. Journal of Environmental Management, 351, 119648.

Häder, D. P., Banaszak, A. T., Villafañe, V. E., Narvarte, M. A., González, R. A., & Helbling, E. W. (2020). Anthropogenic pollution of aquatic ecosystems: Emerging problems with global implications. Science of the Total environment, 713, 136586

Holmlund C.M. and Hammer M. 1999. Ecosystem services generated by fish populations. Ecol. Econ. 29: 253–268.

Leidy, R.A., and Moyle, P.B. (1997). Conservation status of the world's fish fauna: an overview. In Conservation Biology for the Coming Decade. Edited by P.A. Fiedler and P.M. Karieva. Chapman & Hall, New York, New York. pp. 187–227.

Limburg K.E., Hughes R.M., Jackson D.C., and Czech B. 2011. Human population increase, economic growth, and fish conservation: collision course or savvy stewardship? Fisheries. 36: 27–34.

Moore, A. C., and Kumble, S. (2024). Community-based conservation and restoration in coastal wetlands: A review. Wetlands, 44(5), 1-9.

Nash, C., (2010). The history of aquaculture. John Wiley & Sons.

Postel, S., and Carpenter, S. (1997). Freshwater ecosystem services. In Nature's Services: Societal Dependence on Natural Ecosystems. Edited by G.C. Daily. Island Press, Washington, DC. pp. 195–214.

Prasad, S., Wei, Y., Chaminda, T., Ritigala, T., Yu, L., Jinadasa, K. B. S. N., and Zhang, Z. (2024). Spatiotemporal Assessment of Water Pollution for Beira Lake, Sri Lanka. Water, 16(11), 1616.

Raghubanshi, A.S., L.C. Rai, J.P. Gaur and J.S. Singh (2005) Invasive alien species and biodiversity in India, Current Science, 88 (4) : 539-540.

Ramachandra, T.V. (2001). Restoration and Management Strategies of Wetlands in DevelopingCountries, The Greendisk Environmental Journal

Ramachandra, T.V., Ahalya, N. and Rajinikanth. R. (2001). Fish homicide by civic authorities: the latest episode of euthanasia. In Proceedings of Enviro 2001 - National Conference on Control of Industrial Pollution and Environmental Degradation, Sept 14-15, 2001.

Ramachandra, T.V., Kiran R and Ahalya N. (2002) Status, Conservation and Management of Wetlands, Allied Publishers Pvt Ltd, Bangalore.

Rapport D.J., Costanza R., and McMichael A.J. 1998. Assessing ecosystem health. Trends Ecol. Evol. 13: 397–402.

Richter B.D., Braun D.P., Mendelson M.A., and Master L.L. (1997). Threats to imperiled freshwater fauna. Conserv. Biol. 11: 1081–1093.

Srivastava, R. P., Kumar, S. and Tiwari, A. (2024). Continuous emission monitoring systems (CEMS) in India: Performance evaluation, policy gaps and financial implications for effective air pollution control. Journal of Environmental Management, 359, 120584.

Subasinghe, R. (2014). Fishing for development, background paper for session 3: Green growth in fisheries and aquaculture; https://www.oecd.org/tad/events/Fishing-for-development-2014- Session-3-GG-in-aquaculture.pdf.

Tang, S., Zhang, T., Lu, J., Li, D., Pan, J. and Duan, C. 2015. Temporal and spatial variation in fish assemblages in Lake Taihu, China. Journal of Freshwater Ecology, 30(1): pp.181-196.

Tonn, B., English, M., and Travis, C. (2000). A framework for understanding and improving environmental decision making. Journal of Environmental Planning and Management, 43(2), 163-183.

Vörösmarty C.J., McIntyre P.B., Gessner M.O., Dudgeon D., Prusevich A., Green P., Glidden S., Bunn S.E., Sullivan C.A., Reidy Liermann C., and Davies P.M. (2010). Global threats to human water security and river biodiversity. Nature, 467: 555–561.

Vörösmarty, C.J., Lévêque, C., and Revenga, C. (2005). Fresh water. In Millennium Ecosystem Assessment, Volume 1: Conditions and Trends Working Group Report. Edited by R. Bos, C.

Whitfield, A.K. 1999. Ichthyofaunal assemblages in estuaries: a South African case study. Reviews in Fish Biology and Fisheries, 9: pp.151-186.

World Bank, Fish to 2030: prospects for fisheries and aquaculture, 2014. \

Zerbe, S. (2022). A Brief Introduction to Restoration Ecology and Ecosystem Restoration. In Restoration of Multifunctional Cultural Landscapes: Merging Tradition and Innovation for a Sustainable Future (pp. 281-312). Cham: Springer International Publishing

Index